OXFORD MATHEMATICAL MONOGRAPHS

OXFORD MATHEMATICAL MONOGRAPHS

A. Belleni-Morante: *Applied semigroups and evolution equations*
I.G. Macdonald: *Symmetric functions and Hall polynomials*
J.W.P. Hirschfeld: *Projective geometries over finite fields*
N. Woodhouse: *Geometric quantization*
A.M. Arthurs: *Complementary variational principles Second edition*
P.L. Bhatnagar: *Nonlinear waves in one-dimensional dispersive systems*
N. Aronszajn, T.M. Creese, and L.J. Lipkin: *Polyharmonic functions*
J.A. Goldstein: *Semigroups of linear operators*
M. Rosenblum and J. Rovnyak: *Hardy classes and operator theory*
J.W.P. Hirschfeld: *Finite projective spaces of three dimensions*
K. Iwasawa: *Local class field theory*
A. Pressley and G. Segal: *Loop groups*
J.C. Lennox and S.E. Stonehewer: *Subnormal subgroups of groups*
D.E. Edmunds and W.D. Evans: *Spectral theory and differential operators*
Wang Jianhua: *The theory of games*
S. Omatu and J.H. Seinfeld: *Distributed parameter systems: theory and applications*
D. Holt and W. Plesken: *Perfect groups*
J. Hilgert, K.H. Hofmann, and J.D. Lawson: *Lie groups, convex cones, and semigroups*
S. Dineen: *The Schwarz lemma*
B. Dwork: *Generalized hypergeometric functions*
R.J. Baston and M.G. Eastwood: *The Penrose transform: its interaction with representation theory*
S.K. Donaldson and P.B. Kronheimer: *The geometry of four-manifolds*
T. Petrie and J. Randall: *Connections, definite forms, and four-manifolds*
R. Henstock: *The general theory of integration*
D.W. Robinson: *Elliptic operators and Lie groups*
A.G. Werschulz: *The computational complexity of differential and integral equations*
J.B. Griffiths: *Colliding plane waves in general relativity*
L. Evens: *The cohomology of groups*
P.N. Hoffman and J.F. Humphreys: *Projective representations of the symmetric groups*
I. Györi and G. Ladas: *The oscillation theory of delay differential equations*
G. Effinger and D. Hayes: *Additive number theory of polynomials over a finite field*

Additive Number Theory of Polynomials Over a Finite Field

GOVE W. EFFINGER

Skidmore College
Saratoga Springs, New York

and

DAVID R. HAYES

University of Massachusetts
Amherst, Massachusetts

CLARENDON PRESS · OXFORD

This book has been printed digitally and produced in a standard specification in order to ensure its continuing availability

OXFORD
UNIVERSITY PRESS

Great Clarendon Street, Oxford OX2 6DP

Oxford University Press is a department of the University of Oxford.
It furthers the University's objective of excellence in research, scholarship,
and education by publishing worldwide in

Oxford New York

Auckland Cape Town Dar es Salaam Hong Kong Karachi
Kuala Lumpur Madrid Melbourne Mexico City Nairobi
New Delhi Shanghai Taipei Toronto

With offices in

Argentina Austria Brazil Chile Czech Republic France Greece
Guatemala Hungary Italy Japan South Korea Poland Portugal
Singapore Switzerland Thailand Turkey Ukraine Vietnam

Published in the United States
by Oxford University Press Inc., New York

Reprinted 2011

ISBN 978-0-19-853583-6

To Leonard Carlitz, with affection and admiration

and

To the memory of Cecil Effinger, 1914–1990.

Preface

Most of the classic problems of additive number theory can be formulated for the ring $\mathbb{F}_q[T]$ of polynomials over the finite field of q elements. In this monograph, we treat the analogues in $\mathbb{F}_q[T]$ of the Waring problem and the Vinogradov 3-primes problem. Nowadays, one can attack additive problems over $\mathbb{Z}$ and $\mathbb{F}_q[T]$ in a uniform manner by applying the Hardy–Littlewood circle method in its adèlic form (see Weil 1962, 1965 and Mars 1973). For the *global theory*, we have chosen to concentrated our efforts on $\mathbb{F}_q[T]$ for two reasons. First, the difficulties at the archimedean place of $\mathbb{Q}$ mar an otherwise beautiful theoretical development. Second, of course, additive number theory over $\mathbb{Z}$ has already received careful treatments by a number of authors (especially Vaughan 1981), and we would feel hard pressed to offer improvements in either style or content. However, we have arranged our development of the *local theory* at the non-archimedean places so that it applies equally well over either $\mathbb{Z}$ or $\mathbb{F}_q[T]$.

Sometimes problems in the 'function field case' of number theory are easier then their classical analogues. This truism certainly holds for the polynomial 3-primes problem, although the reader who works through all the details in Chapter 7 might be inclined to think otherwise. The reason is that the generalized Riemann hypothesis is valid over the rational function field $\mathbb{F}_q(T)$. We may therefore follow the original Hardy–Littlewood (1922a) attack on this problem without need of the ingenious evasions of Vinogradov. This approach leads to a very good error term in the asymptotic formula and, in fact, to a complete solution of the polynomial 3-primes problem (see the Appendix). However, the Waring problem seems to be more difficult in the polynomial setting because of problems arising from the finite characteristic. For example, the complete elucidation of the Waring problem for squares requires the Riemann hypothesis (see §1.2).

We hope that this monograph will serve as a useful resource for research mathematicians interested in the additive arithmetic of $\mathbb{F}_q[T]$. We have therefore put considerable effort into compiling the bibliography. The reader will find many papers listed there that do not fall within the rather narrow focus of the text. Nonetheless, our exposition is directed toward a first or second year postgraduate student who knows basic number theory and has mastered a standard abstract algebra course. Even though we use the methods of harmonic analysis throughout, we do not really regard this subject as a prerequisite for reading the text. The reader who is willing to accept the Fourier inversion theorem (carefully introduced in §§2.2, 2.3 and 4.3) and to do a bit of reading—say in Loomis (1953)—will find additive number theory to be a useful context for the study of this powerful analytic machinery. We do assume a knowledge of the Riemann hypothesis for smooth, projective curves over $\mathbb{F}_q$; and at one crucial point we assume a little class field theory. The applications of these results are, we hope, clearly stated so that they can be taken on faith by the student who wants to get on with the main thrust of the text and perhaps return to these important topics in a later course of study.

We have tried to follow the familiar Hardy–Littlewood notation as much as possible. For example, we write 'A/Q' for a quotient of polynomials where classically one might expect a Farey fraction 'a/q'. However, we write 'd' for the exponent in the Waring problem because we could not overcome the impulse to reserve 'k' for the field $\mathbb{F}_q(T)$. The notation $\sum'$ always indicates a summation over monic (or, as we prefer to call them, *positive*) polynomials.

We thank M. Car and W.A. Webb for their comments on an early version of Chapter 1. We are very grateful to M. Car for providing us with copies of some of her unpublished work. A suggestion in Descombes (1966–7) was instrumental to our formulation of the Goldbach and 3-primes problems in §1.3. We thank Christopher Hayes for his expert conversion of our typescript into LaTeX.

Finally, we owe special thanks to J.-P. Serre for his encouragement and interest in seeing this work into print.

New York
Amherst
May 1991

G.W.E.
D.R.H.

Contents

List of symbols

Symbol	Meaning	Pages
$\mathbf{A}$, $\mathbf{A}_q$	ring of polynomials $\mathbb{F}_q[T]$	1,21
$\mathbb{A}_k$	adèle ring	55
$\mathbb{A}_k/k$	adèle class group	14,56
a	$(3/4)(r-1)$	107
$B_{q,a}$	parameter of 3-primes analysis	107
c_f	Fourier coefficient of f	28
$c_*(d)$	index in $\mathbf{U}_*$ of d-th powers	45
$C_*(h)$	product of $c_*(d_i)$	48
$\mathbf{C}_k$	idèle class group	72
$\mathbf{C}_k^0$	idèle class group of degree 0	73
C_q	parameter in character sum bound	81-82
$C_i(q,r), i=1..4$	error constants in 3-primes analysis	110-119
$C(q,r)$	sum of error constants	120
D, D_0, D^ϵ	fundamental domain for $\mathbb{A}_k/k$	56,100
D_{MAJOR}, D_{minor}	union of major (resp. minor) arcs	128
dt, d_*t	additive Haar measure	14,25
$d_*^\times u$, $d_*^{(1)}u$	multiplicative Haar measure	37,40
$d(\chi)$	$\deg(\mathfrak{m}(\chi)-2)$	79
$\mathfrak{D}_k$	divisor group of k	72
$\mathfrak{D}_S$	divisors prime to finite set S	76
$\mathfrak{d}$	divisor of k	72
E	additive character of $\mathbb{A}_k$	15,59
E_∞, E_v	local components of E	59
E^P	$E(Pt)$	89
$\mathbb{F}_q$	finite field of order q	1,21
$F_r(t)$	character sum for $\deg P = r$	13,89
$g(d)$	Waring number for degree d	6,11
$G(d)$	almost all Waring number	6,11
$G(\chi,\Lambda)$	Gauss sum	37,90
$G^{(1)}(\chi,\Lambda)$	modified Gauss sum	40
$g(t)$	character sum for Waring problem	124
h	change of variable polynomial	31,43,45

Symbol	Meaning	Pages
$H_r(t)$	character sum for deg $P < r$	13,89
$H_{a,b}(t)$	partial character sum	103
$H^*_{a,b}(t)$	partial sum of Gauss sums	107
$I(M)$	integral for 3-primes (resp. Waring)	108,137
$I_n(M)$	error integral in Waring analysis	137
J_k	idèle group of k	71
J_k^0	kernel of degree map on idèles	73
k	field of fractions of $\mathbf{A}$	14
k_*	complete valued field	23
k_P, k_{v}	completion of k at a place	16,55
k_∞	completion of k at infinity	14
$\mathbf{K}(P)$, $\mathbf{K}(*)$	residue class field	22,23
$L_q(r)$	logarithmic sum	15,103
$L_{q,a}(r)$	partial logarithmic sum	104
$L(s,\chi)$	L-function of χ	76
$\mathbf{M}$	places of k	22
$\mathbf{M}_{\mathrm{fin}}$	finite places of k	16,22
$m(\chi)$	conductor of local character	24
$\mathfrak{m}(\chi)$	conductor of global character	76
M^*	$M/T^{\deg M}$	91
$\mathcal{M}(\chi)$	integral divisors	76
$n(\Lambda)$	conductor of additive character	24
$\mathfrak{n}(t)$	conductor of an adèle	18,93
$N(M)$	number of 3 primes representations	13,89
$N_{d,s}(M)$	number of Waring representations	123
$N_h(t)$	nonzero point count in hypersurface	51
$N(\epsilon)$	parameter associated with $\omega(Q)$	69
$\mathcal{O}_P$, $\mathcal{O}_{\mathrm{v}}$	ring of integers in k_P or k_{v}	16,55
$\mathcal{O}_*$	ring of integers of k_*	23
$\mathcal{O}$	product of $\mathcal{O}_P$ over finite places	16,62
$\mathfrak{p}_{\mathrm{v}}$	prime divisor	72
q_*	order of $\mathbf{K}(*)$	23
$R(\tilde{t})$	finite part of Waring integral	129
$\mathcal{R}$	reduced set of rationals	65
$S(M)$	singular series	15
$S(h,V)$	singular set of h on V	30
$S(\chi)$	set of places dividing $\mathfrak{m}(\chi)$	76
S	$S(\chi) \cup \{\infty\}$	77
$S_d(f,n)$	Weyl sum	131
$S(q,r)$	lower bound for singular series	142

Symbol	**Meaning**	**Pages**
T	indeterminant over F_q	1,21
$t^{(r)}$	adèle with adjusted ∞-place	98
$\mathcal{T}_Q$	Q-toroid	100
$\mathbf{T}$	unit circle in $\mathbb{C}$	13,23
$\mathbf{U}_*$, $\mathbf{U}_P$	units of $\mathcal{O}_*$ or $\mathcal{O}_P$	24
$\mathbf{U}_*^{(\nu)}$	$1+\pi_*^\nu \mathcal{O}_*$	24
v_∞, v_P, v_*	discrete valuation	21-23
v	place of k	55
V	compact subset of $\mathcal{O}^s$	62
W	fundamental domain of $\mathbf{C}_k^0$	71,90
$W_i, i=1,2,3$	decomposition of W	94
α	2 if $q=2$, $3/4$ otherwise	117
$\delta(t_\infty)$	characteristic function	98
Δ_A	polynomial operator	131
$\theta_*(t;h,V)$, $\theta_P(t;h,V)$	local Radon-Nikodym derivative	30
$\theta_*^u(t;h,V)$	unnormalized derivative	32
$\theta(t)$, $\theta(t;h,V)$	global Radon-Nikodym derivative	16,63
Λ	additive character of k_*^+	24
$\Lambda^a(x)$	$\Lambda(ax)$	24,58
$\mu(Q)$	Möbius function	66
$\mu(\mathfrak{d})$	Möbius function on divisors	93
μ, μ_2	parameters in Waring analysis	138
π_*	uniformizing parameter	23
$\pi_\nu(\chi)$	character sum	79
$\varpi_q(r)$	count of irreducible polynomials	116
ρ_{v}	product of local measures	62
$\tau(L)$	number of positive divisors of L	131
$\Phi(Q)$	Euler function	66
$\Phi(\mathfrak{d})$	Euler function on divisors	93
χ	multiplicative character	24
χ_*	multiplicative character of $\mathfrak{D}_S$	76,91
$\psi(n)$	positive irreducibles of degree n	92
$\omega(Q)$	number of positive irreducible divisors of Q	69
$\bar{\omega}(r)$	maximum $\omega(Q)$ over deg $Q=r$	69
$\{x\}$	least integer function	4
$\|x\|$	absolute value	12,21
$\|x\|_*$, $\|x\|_P$	normalized absolute value	22,23

Symbol	**Meaning**	**Pages**
$\|\mathfrak{d}\|$	absolute value of a divisor	73
$\tilde{E}$, $\tilde{t}$, etc.	restriction to finite adèles	15,56,93
$\hat{G}$, $\hat{k}$, etc.	character group	23
$\hat{f}$, $\hat{\theta}$, etc.	Fourier transform	26

1

The polynomial Waring and Goldbach problems

From 1930 to 1932, Leonard Carlitz held a National Research Fellowship. He spent a substantial portion of his fellowship at Cambridge University, where he met Raymond Paley, a brilliant young English mathematician of about the same age as himself. In the decade just gone by, Hardy and Littlewood had captured the interest of the mathematical world with a remarkable series of papers on two classic problems in additive number theory—the *Waring problem* and the *Goldbach problem.* The new technique that they relentlessly focused on these problems was the *circle method*, first introduced by Hardy and Ramanujan in 1918 to derive an asymptotic expansion for the partition function.

Perhaps it was his mentor, Littlewood, who influenced Paley to work out the Waring problem for the ring $\mathbf{A} = \mathbf{A}_q = \mathbb{F}_q[T]$ of polynomials over a finite field. Paley, who was a very talented analyst, is best known for his work with Wiener on Fourier transforms (Paley and Wiener 1934). The main result of Paley's investigations into the polynomial Waring problem may be stated as follows:

Theorem 1.1. (Paley 1933) *Let d be prime to the characteristic of $\mathbf{A}$. Then all polynomials in $\mathbf{A}$ which are not barred by congruences from being the sum of d-th powers can be expressed as the sum of the d-th powers of at most s polynomials in $\mathbf{A}$, where $s = s(d)$ is a number which depends only on d.*

The circle method is an analytic technique, but Paley used purely algebraic methods in his proof. After all, where in characteristic $p > 0$ might one find a proper analogue for the unit circle? A modern mathematician might think of introducing Haar measure on some compact group appropriate to the characteristic p universe. Such ideas were not unknown when Paley proved this theorem, but they were not part of the standard toolbox of the analyst. In fact, Haar's paper proving the general existence theorem for invariant measures on locally compact groups appeared also in 1933.

Carlitz was delighted to find someone at Cambridge who shared his interest in the arithmetic of polynomials over a finite field. Much as he admired Paley's theorem, Carlitz understood that Paley had proved an analogue not of the classic Waring problem but rather of what is known as the 'easier' Waring problem. In Theorem 1.1, the representation of a given polynomial M as a sum of d-th powers might involve much cancellation among monomials T^n with $n > \deg M$. Carlitz therefore formulated a Waring problem for $\mathbf{A}$ (see §1.1 below) which limits the degrees of the summands; and he began to think about the case $d = 2$ (Carlitz 1933, 1935, 1937). Later, his student Eckford Cohen gave a definitive treatment (1947a,b, 1948) to representations by sums of squares in $\mathbf{A}$. The methods used in all these $d = 2$ papers are combinatorial and algebraic.

Because of his tragic death in a skiing accident (see Wiener 1933), we can never know if Paley might someday have returned to his work with polynomials over $\mathbb{F}_q$. As an analyst interested in Fourier transforms, he would certainly have become well acquainted with Haar measure. One cannot doubt that Paley would have understood the possibility of applying this new tool to the additive number theory of the ring $\mathbf{A}$. A major aim of this book is to explore this possibility in some depth.

1.1 The polynomial Waring problem

The reader is probably familiar with congruence conditions which can obstruct the representation of a positive integer as the sum of a fixed number of d-th powers. For example, a prime number of the form $4r + 3$ is not the sum of two squares. However, as Waring surmised, for any given d, there always exists an s such that every positive integer is a sum of at most s d-th powers. Theorem 1.1 suggests that congruence obstructions to representations by d-th powers in $\mathbf{A}$ do not necessarily disappear when the number of summands becomes large. We will determine the precise nature of these persistent obstructions. Before starting work on this task, we pause to introduce an unconventional term for monic polynomials. In this book, we refer to such polynomials as *positive*. We hope the reader will eventually appreciate the reasons for this deviation from the standard word. In the older literature, positive polynomials were called *primary*, a word which, unlike 'monic', is not neutral.

Let $d \geq 1$ be an exponent which is not divisible by the characteristic of $\mathbf{A}$. Suppose we wish to write a polynomial $M \in \mathbf{A}$ in the form

$$M = X_1^d + X_2^d + \cdots + X_s^d \tag{1.1}$$

with $X_1, X_2, \ldots, X_s \in \mathbf{A}$. If (1.1) has no solutions in the quotient ring $\mathbf{A}/H\mathbf{A}$ for some given positive modulus $H \in \mathbf{A}$, then (by the Chinese remainder theorem) it has no solution in $\mathbf{A}/P^n\mathbf{A}$ for some $P^n|H$, where P is a positive irreducible. Since d is not divisible by the characteristic of $\mathbf{A}$, any non-zero solution of (1.1) in $\mathbf{A}/P\mathbf{A}$ can be lifted to $\mathbf{A}/P^n\mathbf{A}$ by Hensel's lemma. Now for $s > d$, if (1.1) has *any* solution in $\mathbf{A}/P\mathbf{A}$, then it will have a non-zero solution by the theorem of Chevalley (cf. Chapter 6 of Lidl and Niederreiter 1983). We may therefore restrict our search for persistent congruence obstructions of (1.1) to positive irreducible moduli.

Definition 1.2. *Let $q = p^e$, where p is the characteristic of $\mathbf{F}_q$. An integer N is called a q-*norm *if $N = (q-1)/(p^a-1)$ for some divisor a of e other than e itself.*

Lemma 1.3. *Suppose $d = mN$ is a multiple of a q^b-norm N for some $b \geq 1$. Then for any positive irreducible P of degree b, there exist polynomials M for which* (1.1) *is not solvable modulo P.*

Proof. For $X \in \mathbf{A}$, X^N modulo P is the norm of X down to the subfield of $\mathbf{A}/P\mathbf{A}$ which has p^a elements. If M does not belong to this subfield modulo P, then (1.1) has no solution in $\mathbf{A}/P\mathbf{A}$. ∎

Definition 1.4. *An exponent d is said to be q-*proper *if d is not divisible by any q-norm.*

Lemma 1.5. *For a q-proper exponent d, let $\delta = \mathbf{GCD}(d, q-1)$. Then every element of $\mathbf{F}_q$ is the sum of at most δ d-th powers of elements of $\mathbf{F}_q$.*

Proof. Following Tornheim (1938) (but see also Schwarz 1948a,b), we observe that the elements of $\mathbf{F}_q$ which are sums of d-th powers form a subfield E of $\mathbf{F}_q$. If N is the index of $E^\times$ in $\mathbf{F}_q^\times$, then N divides d because $\mathbf{F}_q^\times$ is cyclic and $\gamma^d \in E^\times$ for all $\gamma \in \mathbf{F}_q^\times$. Therefore $N = 1$, since otherwise N is a q-norm.

For each integer $i \geq 1$, let Σ_i be the set of elements of $\mathbf{F}_q$ which are sums of at most i d-th powers. Let s be the first index i such that $\Sigma_i = \mathbf{F}_q$. Then we must have $\Sigma_1 \subsetneq \Sigma_2 \subsetneq \cdots \subsetneq \Sigma_s$. To prove it, let

$$\gamma = \alpha_1^d + \alpha_2^d + \cdots + \alpha_s^d$$

be an element of Σ_s which does not belong to Σ_{s-1}. Then $\sum_{j \leq i} \alpha_j^d$ belongs to Σ_i but not to Σ_{i-1}.

We may now show that $s \leq \delta$. Let $G \subseteq \mathbf{F}_q^\times$ be the subgroup of d-th powers. If $\gamma \in \Sigma_i$, $\gamma \neq 0$, then clearly the whole coset $\gamma G \subseteq \Sigma_i$. Therefore, each Σ_i includes at least one coset of G which is not in Σ_{i-1}. It follows that the index of G in $\mathbf{F}_q^\times$ is greater than or equal to s. That index is δ. ∎

Corollary 1.6. *For a q-proper exponent d, let $\delta = \mathbf{GCD}(d, q-1)$. Let $\gamma_1, \gamma_2, \ldots, \gamma_s \in \mathbb{F}_q^\times$. If $s > \max\{1, \delta^2 - \delta\}$, then the equation*

$$\gamma_1 X_1^d + \gamma_2 X_2^d + \cdots + \gamma_s X_s^d = \gamma \tag{1.2}$$

has a non-zero solution in $\mathbb{F}_q^s$ for every $\gamma \in \mathbb{F}_q$.

Proof. Since $s > \delta^2 - \delta$, at least δ of the coefficients $\gamma_1, \gamma_2, \ldots, \gamma_s$ must belong to the same coset of the subgroup of d-th powers in $\mathbb{F}_q^\times$. We conclude then from Lemma 1.5 that (1.2) has a solution in $\mathbb{F}_q^s$. Because $s > \delta$, the theorem of Chevalley guarantees a non-zero solution of this equation. ■

The analysis above shows that congruence obstructions to the representation (1.1) for large s occur if and only if d is a multiple of a q^b-norm for some $b \geq 1$. We may therefore reformulate the major thrust of Paley's theorem as follows:

Theorem 1.7. (Easier Polynomial Waring) *Suppose d is prime to the characteristic of $\mathbf{A}_q$ and q^b-proper for every $b \geq 1$. Then there exist positive integers n and s such that every polynomial $M \in \mathbf{A}_q$ of degree r has a representation* (1.1) *by polynomials $X_1, X_2, \ldots, X_s$ each of degree less than or equal to nr.*

Proof. By Theorem 1.1, there is a representation

$$T = Y_1^d + Y_2^d + \cdots + Y_s^d \tag{1.3}$$

of T as a sum of d-th powers in $\mathbf{A}$. Substituting M for T in (1.3), we see that the theorem is valid with $n = \max\{\deg Y_i : 1 \leq i \leq s\}$. ■

Tornheim (1938) has shown that if $d < \mathrm{char}(\mathbf{A})$, then (1.3) holds with $s \leq d(d+1)$.

As Carlitz observed, this last theorem is not a solution to the strict analogue in $\mathbf{A}$ of the classical Waring problem. In such an analogue, the degrees of the summands must be restricted as much as possible.

Definition 1.8. *For a real number x, let $\{x\}$ be the least integer which is greater than or equal to x. We say that a polynomial $M \in \mathbf{A}$ is the* strict sum of s d-th powers *if it has a representation* (1.1) *by polynomials $X_1, X_2, \ldots, X_s$ which are each of degree less than or equal to $\{\deg M/d\}$.*

The (strict) polynomial Waring problem is formulated in the following theorem.

Theorem 1.9. (Polynomial Waring) *Suppose $0 < d < p = \mathrm{char}(\mathbf{A}_q)$. Then there exists a positive integer $s = s(d)$, independent of q, such that every $M \in \mathbf{A}_q$ is the strict sum of s d-th powers.*

Since $N \geq 1 + \sqrt{q}$ for any q-norm N, any $d < p$ will be q^b-proper for every $b \geq 1$. Therefore, there can be no congruence obstructions to Theorem 1.9.

The condition $d < p$ is essential, as one can see by considering strict representations of polynomials M of degree one. At first, this condition appears to be rather severe. However, suppose we wish to study sums of, say, seventh powers in $\mathbf{A}_q$. For $p > 7$, we are assured by Theorem 1.9 that every polynomial is the strict sum of seventh powers with a value for s which is independent of q. We may then proceed to develop more precise information about seventh powers. For example, we may try to find an explicit bound for the minimum value of s which suffices for all $\mathbf{A}_q$ with $p > 7$. Here we encounter a principle which is worth the effort of a formal statement.

Maxim 1.10. *In the additive number theory of polynomials, one should always consider the universe of all the rings $\mathbf{A}_q$ as the primary object.*

In Chapter 8, we prove an asymptotic result which shows that when $d < p$, (1.1) holds with $s = d^2 2^d$ for polynomials $X_1, X_2, \ldots, X_s$, each of degree less than or equal to $\{\deg M/d\}$, whenever $q^{\deg M}$ is larger than an absolute constant. Thus, there can be only finitely many polynomials M in the universe of all the $\mathbf{A}_q$ which are exceptions to Theorem 1.9. In the classical Waring problem, a finite number of exceptions is easy to deal with because a positive integer n is the sum of n ones. The polynomial exceptions are more of a challenge.

Lemma 1.11. (Webb 1973a) *For $n \geq 0$, let $V_n \subset \mathbf{A}_q$ be the $(nd+1)$-dimensional $\mathbb{F}_q$-subspace of all polynomials of degree less than or equal to nd. If $d < p$, then V_n has a basis consisting of d-th powers.*

Proof. Since the constant polynomial 1 is a basis for V_0, it suffices if we prove that V_n/V_{n-1} has a basis of d-th powers for all $n \geq 1$. Let $\{\alpha_1, \ldots, \alpha_d\}$ be any set of d distinct elements of $\mathbb{F}_q$. In order to show that the d-th powers

$$(T + \alpha_i)^d = \sum_{j=0}^{d} \binom{d}{j} T^{d-j} \alpha_i^j$$

for $1 \leq i \leq d$ are independent over $\mathbb{F}_q$, it is enough to show that the $d \times d$ matrix

$$\mathcal{M} = \left(\tbinom{d}{j}\alpha_i^j\right)_{1 \leq i \leq d, 1 \leq j \leq d}$$

has non-zero determinant. As we easily compute,

$$\det(\mathcal{M}) = \prod_{i<j}(\alpha_j - \alpha_i) \cdot \prod_{j=0}^{d} \binom{d}{j},$$

which is non-zero because p does not divide any of the binomial coefficients in the product on the right hand side. Now the d-th powers

$$T^{(n-1)d}(T+\alpha_i)^d \tag{1.4}$$

for $1 \leq i \leq d$ are also independent in V_n/V_{n-1} because the determinant $\det(\mathcal{M})$ remains non-zero. As the dimension of V_n/V_{n-1} is d, the d-th powers (1.4) form a basis for this quotient. ■

This lemma together with Lemma 1.5 proves the following.

Lemma 1.12. *If $d < p$, then every $M \in V_n$ is the sum of not more than $d(nd+1)$ d-th powers in V_n.*

Just as in the classical Waring problem, determining the minimal value of s for any given d is a difficult problem which is deeply tied to the arithmetic properties of the rings $\mathbf{A}_q$. Theorem 1.9 justifies the following definition.

Definition 1.13. *Given $d \geq 2$, let $g(d)$ be the smallest value of s such that for every q with $p = \mathrm{char}(\mathbf{A}_q) > d$, every $M \in \mathbf{A}_q$ is the strict sum of s d-th powers in $\mathbf{A}_q$. Let $G(d)$ be the smallest s such that this same condition holds except possibly for a finite number of polynomials in the universe of all the $\mathbf{A}_q$ with $p > d$.*

It follows from Theorem 15 of Cohen (1947a) that $g(2) \leq 4$. Since representation by sums of two squares is well understood (Leahey 1967), $G(2) > 2$. In the next section, we will prove a theorem of Serre which shows that $G(2) = 3$.

Theorem 1.9 was proved independently by Car (1971a), Webb (1973a) and Kubota (1974). Their proofs all employ a polynomial analogue of the circle method. Previously, Carlitz had proved a version of Theorem 1.9 using the algebraic methods described in his papers (Carlitz 1947b, 1948). Unfortunately, this proof was never published.

Exercises 1.1

1. (Carlitz 1975) Consider the easier Waring problem with $d = 2$. Show that for odd q, every polynomial $M \in \mathbf{A}_q$ is the sum of three squares of polynomials of degree less than or equal to $\deg M$.

2. For odd q, the quadratic forms

$$x_1^2 + x_2^2 + x_3^2 + x_4^2 \qquad \text{and} \qquad x_1x_2 + x_3x_4$$

are equivalent over $\mathbb{F}_q$. This means that there is a non-singular, homogeneous $\mathbb{F}_q$-linear change of variables which transforms one form into the other. Prove that, for odd q, every $M \in \mathbf{A}_q$ is the strict sum of four squares.

1.2 Sums of squares

Throughout this section, q is assumed to be odd. Cohen (1948) proved an explicit formula for the number of representations of a given polynomial $M \in \mathbf{A}_q$ as the strict sum of three squares. Unfortunately, it is not at all clear from this formula that the number of such representations is non-zero. While it is easy to show that every $M \in \mathbf{A}_q$ is the strict sum of at most four squares (see Exercise 1.1(2)), the three squares problem is decidedly non-trivial. Using the Riemann hypothesis for smooth projective curves over $\mathbb{F}_q$ (Weil 1948), Serre proved the following theorem circa 1982.

Theorem 1.14. (Serre) *Let q be odd. Except for polynomials M of degrees three and four in $\mathbf{A}_3$, every $M \in \mathbf{A}_q$ is the strict sum of three squares.*

We present Serre's proof in the remainder of this section. This is the first time that the proof has appeared in print. The authors are grateful to Serre for kindly giving us permission to include his theorem and proof in this monograph.

In proving Theorem 1.14, we will assume that M is square-free. Clearly, this assumption entails no loss of generality. We first prove two lemmas.

Lemma 1.15. *The quadratic forms $x^2+y^2+z^2$ and $-y^2+xz$ are equivalent over $\mathbb{F}_q$.*

Proof. The form $x^2+y^2+z^2$ has discriminant 1, and the form $-y^2+xz$ has discriminant 1/4. By Chapter III, §6 of Artin (1957), the quadratic character of the discriminant characterizes the class of the form. Therefore there is a non-singular, linear substitution

$$\begin{aligned} x &\leftarrow \alpha_1 x + \beta_1 y + \gamma_1 z \\ y &\leftarrow \alpha_2 x + \beta_2 y + \gamma_2 z \\ z &\leftarrow \alpha_3 x + \beta_3 y + \gamma_3 z \end{aligned} \tag{1.5}$$

defined over $\mathbb{F}_q$ which transforms $x^2+y^2+z^2$ into $-y^2+xz$. ■

We may interpret this lemma as follows: if x, y, z, and w are quantities in any algebra $\mathcal{A}$ over $\mathbb{F}_q$ such that $y^2 - xz = w$, then $-w$ is a sum of three squares in $\mathcal{A}$.

Lemma 1.16. *Suppose $\deg M \leq 2n$, where $n \geq 1$. Suppose further that there is a divisor d of n and distinct positive irreducibles $P_1, \ldots, P_{n/d}$, each of degree d, such that the congruence $y^2 \equiv M \pmod{P_i}$ is solvable for $1 \leq i \leq n/d$. Then $-M$ is the strict sum of three squares.*

Proof. Let $Z = \prod P_i$, a polynomial of degree n. By the Chinese remainder theorem, the congruence $y^2 \equiv M \pmod{Z}$ has a solution Y with $\deg Y < n$. Therefore, $Y^2 - M = XZ$ for some polynomial X. Now $\deg X \leq n$ because $\deg Z = n$ and the left hand side of this equality has degree $2n$ or less. Applying the change of variables (1.5) with $x = X$, $y = Y$, and $z = Z$, we obtain a representation of $-M$ as a strict sum of three squares. ∎

For the remainder of the proof, M will be a fixed square-free polynomial in $\mathbf{A}_q$ of degree $m \leq 2n$. The idea now is to translate the hypothesis of Lemma 1.16 into a statement about the affine curve $\mathbf{Y}/\mathbb{F}_q$ defined by the equation $y^2 = M(x)$ and then apply the Riemann hypothesis. Let $\overline{\mathbb{F}}_q$ be the algebraic closure of $\mathbb{F}_q$. If $\alpha \in \overline{\mathbb{F}}_q$, then we let '$\deg \alpha$' denote the degree of the minimal polynomial P_α of α over $\mathbb{F}_q$. The map $\gamma : (\alpha, \beta) \mapsto \deg \alpha$ induces a fibration on the set of points $\mathbf{Y}(\mathbb{F}_{q^n})$.

Definition 1.17. *Let $d|n$. We put*

$$N_d = \#\gamma^{-1}(d) = 2p_d + z_d \tag{1.6}$$

where z_d is the number of points $(\alpha, \beta) \in \mathbf{Y}(\mathbb{F}_{q^n})$ with $\deg \alpha = d$ and $\beta = M(\alpha) = 0$. Let Γ_d be the set of x-coordinates of the points in $\gamma^{-1}(d)$.

As the reader will easily verify, p_d is an integer and $p_d + z_d$ is the cardinality of Γ_d.

Proposition 1.18. *Let d be a divisor of n such that n/d is odd. If $\#\Gamma_d \geq n$, then $-M$ is the strict sum of three squares in $\mathbf{A}_q$.*

Proof. For a given $\alpha \in \Gamma_d$, choose β so that $(\alpha, \beta) \in \mathbf{Y}(\mathbb{F}_{q^n})$. Because n/d is odd, $\mathbb{F}_{q^n}/\mathbb{F}_{q^d}$ contains no quadratic subextensions, and therefore $\beta \in \mathbb{F}_{q^d}$ also. Let $S_\alpha : \mathbf{A}_q \to \mathbb{F}_{q^d}$ be the substitution map determined by $T \mapsto \alpha$, and let $Y \in \mathbf{A}_q$ be a polynomial such that $S_\alpha(Y) = \beta$. Then $S_\alpha(Y^2 - M) = 0 \Rightarrow Y^2 \equiv M \pmod{P_\alpha}$.

Now Γ_d is obviously closed under the Galois action over $\mathbb{F}_q$ and therefore contains all the $\mathbb{F}_q$-conjugates of any one of its elements. The number of distinct minimal polynomials P_α for $\alpha \in \Gamma_d$ is therefore $\#\Gamma_d/d \geq n/d$ by hypothesis. The conclusion now follows by Lemma 1.16. ∎

Let $\mathbf{X}$ be the projective closure of $\mathbf{Y}$, and let $\widetilde{\mathbf{X}}$ be the normalization of $\mathbf{X}$. The curve $\widetilde{\mathbf{X}}$ is a hyperelliptic curve over $\mathbb{F}_q$ of genus $g(\widetilde{\mathbf{X}})$ given by

$$g(\widetilde{\mathbf{X}}) = \begin{cases} (m-1)/2 & \text{if } m \text{ is odd} \\ (m-2)/2 & \text{if } m \text{ is even} \end{cases}$$

(see Chapter III, §5 of Shafarevich 1977). In either case, we have

$$g(\widetilde{\mathbf{X}}) \leq n-1$$

because $m \leq 2n$. By the Riemann hypothesis for $\widetilde{\mathbf{X}}$,

$$\left|q^n + 1 - \#\widetilde{\mathbf{X}}(\mathsf{F}_{q^n})\right| \leq 2g(\widetilde{\mathbf{X}}) \cdot q^{n/2} \leq 2(n-1)q^{n/2}.$$

Now, the natural projection $\widetilde{\mathbf{X}} \to \mathbf{X}$ is injective except over the point at infinity on $\mathbf{X}$, and the inverse image of the point at infinity contains at most two points. Therefore

$$\#\mathbf{Y}(\mathsf{F}_{q^n}) \geq \#\widetilde{\mathbf{X}}(\mathsf{F}_{q^n}) - 2 \geq q^n - 2(n-1)q^{n/2} - 1. \tag{1.7}$$

After Proposition 1.18, we are interested in counting the points in Γ_d for divisors d of n such that n/d is odd. Now by (1.6) and (1.7)

$$\begin{aligned}\sum_{d|n} \#\Gamma_d &= \sum_{d|n} p_d + z_d = \frac{1}{2}\sum_{d|n}(2p_d + z_d) + z_d \\ &= \frac{1}{2}\left(\sum_{d|n} N_d + z_M\right) = \frac{1}{2}(\#\mathbf{Y}(\mathsf{F}_{q^n}) + z_M) \qquad (1.8)\\ &\geq \frac{1}{2}q^n - (n-1)q^{n/2} + \frac{1}{2}(z_M - 1)\end{aligned}$$

where z_M is the number of zeros of $M(x)$ of degree equal to some divisor of n. Let $n = rt$ with $r = 2^\ell$ and t odd. If $d = 2^\nu\delta$ with $0 \leq \nu < \ell$, then $p_d + z_d$ is just the total number of elements in $\overline{\mathsf{F}}_q$ of degree d because then F_{q^n} contains the square root of every element of F_{q^d}. Therefore if $\ell > 0$, then

$$\sum_{\nu=0}^{\ell-1}\sum_{\delta|t} \#\Gamma_{2^\nu\delta} = q^{n/2}.$$

In any case, we conclude from this equation and (1.8) that

$$\sum_{\delta|t} \#\Gamma_{r\delta} \geq \frac{1}{2}q^n - (n-1+e_n)q^{n/2} + \frac{1}{2}(z_M - 1) \tag{1.9}$$

where $e_n = 1$ if n is even and $e_n = 0$ if n is odd.

Table 1.1. Inequality (1.11) with $q = 3$

n	t	$n\tau(t) - r\sigma(t)$	$(3^n - 1)/2 - (n - 1 + e_n)3^{n/2}$
1	1	0	1
2	1	0	−2
3	3	2	2.61
4	1	0	4
5	5	4	58.56

Suppose now that $\#\Gamma_{r\delta} < n$ for every divisor δ of t. Since $r\delta$ divides $\#\Gamma_{r\delta}$, this inequality implies that

$$\#\Gamma_{r\delta} \leq n - r\delta$$

which in turn implies that

$$\sum_{\delta|t} \#\Gamma_{r\delta} \leq n\tau(t) - r\sigma(t) \tag{1.10}$$

where $\tau(t)$ is the number of divisors of t and $\sigma(t)$ is the sum of the divisors of t. Therefore, if the hypothesis of Proposition 1.18 is never valid for any $d|n$, then we must have

$$n\tau(t) - r\sigma(t) \geq (q^n - 1)/2 - (n - 1 + e_n)q^{n/2} \tag{1.11}$$

by (1.9) as $z_M \geq 0$. Table 1.1 tabulates the two sides of this inequality with $q = 3$ and $1 \leq n \leq 5$. For fixed n, the right hand side of (1.11) is an increasing function of $q \geq 3$ whenever its value at $q = 3$ is positive. Therefore, this inequality *fails* for $1 \leq n \leq 5$ and all odd q except when $n = 2$. For $n = 2$ and $q = 5$, the right hand side of (1.11) equals 2. Therefore, the inequality also fails for $n = 2$ and all $q \geq 5$. We may conclude that, except for the case $n = 2$ and $q = 3$, every polynomial $M \in \mathbf{A}_q$ of degree 10 or less is the strict sum of three squares.

In order to finish the proof, it suffices to show that (1.11) fails for $q = 3$ and $n \geq 6$. Since, trivially, $\tau(t) \leq \sqrt{t} + 1$ and $\sigma(t) \geq t$, the left hand side of (1.11) is bounded above by $n(\sqrt{t} + 1) - rt = n\sqrt{t} \leq n^{3/2}$. By the usual methods, one can verify that the function

$$(3^x - 1)/2 - x3^{x/2} - x^{3/2}$$

is strictly positive for all $x \geq 5$. This completes the proof of Theorem 1.14.

The exception $n = 2$ and $q = 3$ is not a fault of the proof. Computations by Webb (1973c) show that there are exactly two polynomials M of degree three and six polynomials M of degree four in $\mathbf{A}_3$ which are *not* the strict sum of three squares. We conclude therefore that $g(2) = 4$ and $G(2) = 3$.

The precise values of $g(d)$ and $G(d)$ are not known at this writing for any $d > 2$. However, good bounds for $G(d)$ have been proved by Car (1972) using the method of Vinogradov.

Exercises 1.2

1. (Serre) Consider the easier Waring problem for $d = 3$. Let F be any field of characteristic not 3. A projective point $P = (x, y, z)$ on the Fermat curve $x^3 + y^3 + z^3 = 0$ is *trivial* if $xyz = 0$. Show that this curve has a non-trivial F-rational point if and only if there are elements $a_i, b_i \in F$, $1 \leq i \leq 3$, such that

$$T = (a_1T + b_1)^3 + (a_2T + b_2)^3 + (a_3T + b_3)^3$$

in $F[T]$. [Hint: Let $P = (a_1, a_2, a_3)$ be a non-trivial F-rational point on the Fermat curve, and let $Q = (b_1, b_2, b_3)$ be the point where the tangent line at P intersects the curve again.]

2. (Serre) For q not a power of 3, the Fermat curve $x^3 + y^3 + z^3 = 0$ has a non-trivial $\mathbb{F}_q$-rational point except when $q = 2, 4, 7, 13, 16$. [Hint: Use the Riemann hypothesis and the fact that the Fermat curve has genus $g = 1$.]

3. Since 3 is a 4-norm, T is not a sum of three cubes in $\mathbb{F}_q[T]$ for $q = 2$ and $q = 4$. Vaserstein (1991) found the identities

$$T = (T^6 - 3T^4 - T^3 - T + 1)^3 + (-T^6 + 3T^4 - T^3 - T - 1)^3 + (-T^5 - 3T^3 + T)^3$$

in $\mathbb{F}_7[T]$ and

$$T = (5T^5 - 3T^4 - 5T^3 + T^2 + 4T + 1)^3 + (T^4 + 6T^3 + 5T^2 - 4T)^3 + (-5T^5 + 3T^4 + 5T^3 - 5T^2 + 5T - 1)^3$$

in $\mathbb{F}_{13}[T]$. Use these results together with Exercises 1 and 2 above to prove the following remark of Serre: let R be a ring with one, and let t be any element of R such that $nt = 0$ for some integer n prime to 2 and 3. Then t is the sum of three cubes in R. [Hint: Using induction on m, show first that the element $T \in (\mathbb{Z}/p^m\mathbb{Z})[T]$, $p \neq 2$ or 3, is the sum of three cubes.]

4. (Research problems) (i) Is T the sum of three cubes in $\mathbb{F}_{16}[T]$? (ii) Is T the sum of three cubes in $\mathbb{Q}[T]$?

1.3 The polynomial Goldbach problem

For which positive $M \in \mathbf{A}$ can we find positive irreducibles P_1 and P_2 in $\mathbf{A}$ with $\deg P_1 = \deg M$ and $\deg P_2 < \deg M$ such that $M = P_1 + P_2$? We are unable to answer this deep question. However, it is easy to see that all linear polynomials and all quadratic polynomials of the form $T^2 + T + \alpha$ in characteristic 2 are too small to have such a representation. There is also at least one congruence obstruction.

Definition 1.19. *For $M \in \mathbf{A}$, the* absolute value *of M is the integer*

$$|M| = q^{\deg M}. \tag{1.12}$$

We say that M is even *if M is divisible by an irreducible polynomial P with $|P| = 2$. Otherwise, we say that M is* odd.

As the reader will easily note, even polynomials can occur only in $\mathbf{A}_2 = \mathbb{F}_2[T]$. Suppose $M \in \mathbf{A}_2$ is the sum of irreducibles P_1 and P_2 with $\deg P_2 < \deg M$. Substituting first $T = 0$ and then $T = 1$ into the identity $M = P_1 + P_2$, we conclude that M is divisible by both T and $T + 1$ unless P_2 has degree one. If $P_2 = T$, then $T + 1$ must divide M; and if $P_2 = T + 1$, then T must divide M. In any case, M must be *even*.

Conjecture 1.20. (Polynomial Goldbach) *For every positive $M \in \mathbf{A}$ with $\deg M \geq 2$, there exist positive irreducibles P_1 and P_2 in $\mathbf{A}$ such that*

$$M = P_1 + P_2 \qquad (\deg P_1 = \deg M, \quad \deg P_2 < \deg M) \tag{1.13}$$

except for polynomials $M \in \mathbf{A}_2$ which are not multiples of $T(T+1)$ and quadratic polynomials $M = T^2 + T + \alpha$ in characteristic 2.

The polynomial Goldbach problem is just as resistant to currently available techniques as is the classical Goldbach problem. See Cherly (1978) for an attempt by sieve methods. However, if we consider *three* irreducible summands instead of two, then we can prove an analogue of Vinogradov's famous 3-primes theorem: *Every sufficiently large odd integer is the sum of three primes.*

Definition 1.21. *A positive polynomial $M \in \mathbf{A}$ with $\deg M = r$ is said to be a* 3-primes polynomial *if it can be written as the sum of three positive irreducibles in $\mathbf{A}$, one of degree r and the other two of degree less than r.*

In Chapter 7, we apply the polynomial circle method to prove the following.

Asymptotic Theorem 1.22. *For every degree $r \geq 5$, there exists a q_r such that if $q \geq q_r$, then every odd, positive $M \in \mathbf{A}_q$ of degree r is a 3-primes polynomial. Further, the sequence $\{q_r\}_{r\geq 5}$ is decreasing, and $q_r = 2$ for all sufficiently large r.*

In proving this theorem, we follow the original Hardy and Littlewood (1922a) attack on the classical 3-primes problem. Their method requires the Riemann hypothesis for *all* the Dirichlet L-functions. Now by Weil (1948), the Riemann hypothesis for smooth, projective curves over $\mathbb{F}_q$ is valid; and we can use Weil's result to establish the Riemann hypothesis for the *L-functions of Dirichlet type* which we introduce in Chapter 5. This result allows us to prove Theorem 1.22 with fantastically small values for the lower bounds q_r (see Table A.1 in Appendix A). Although we do not carry out the details in this monograph, these good bounds together with considerably more work enable us to prove the following.

Theorem 1.23. (Polynomial 3-Primes) *Every odd, positive $M \in \mathbf{A}$ with $\deg M \geq 2$ (except for the case $M = T^2 + \alpha$ in characteristic 2) is a 3-primes polynomial.*

See Appendix A for a discussion of the proof of this theorem.

One can also attack the polynomial 3-primes problem by applying the Vinogradov method for estimating exponential sums. See Car (1971b) for this approach, which avoids an appeal to the Riemann hypothesis.

1.4 The circle method

Let $\mathcal{T}$ be any compact, divisible $\mathbf{A}$-module, and let $E : \mathcal{T} \to \mathbf{T}$ be a non-trivial character of the additive group of $\mathcal{T}$, where $\mathbf{T}$ is the unit circle. We can develop a 'circle method' formalism based on $\mathcal{T}$ for attacking additive problems in $\mathbf{A}$. Consider, for example, the polynomial 3-primes problem (Theorem 1.23). For $S \in \mathbf{A}$ positive of degree $r \geq 2$, let $N(S)$ be the number of representations $S = P_1 + P_2 + P_3$, where P_1, P_2, and P_3 are positive irreducibles such that $\deg P_1 = r$, $\deg P_2 < r$, and $\deg P_3 < r$. For $t \in \mathcal{T}$, we define

$$F_r(t) = \sum_{\deg P = r}{}' E(Pt) \tag{1.14}$$

and

$$H_r(t) = \sum_{\deg P < r}{}' E(Pt) = \sum_{\nu=1}^{r-1} F_\nu(t) \tag{1.15}$$

where P runs first through the positive irreducibles of degree r and then through the positive irreducibles of degree less than r. We have then

$$F_r(t) \cdot H_r(t)^2 = \sum_{\deg S = r}' N(S)\,E(St)\,, \tag{1.16}$$

S running through all positive polynomials of degree r. For a fixed positive M of degree r, the character $t \mapsto E((S-M)t)$ is non-trivial when $S \neq M$ because $\mathcal{T}$ is divisible. We may therefore solve for the coefficient $N(M)$ in (1.16) by multiplying both sides by $E(-Mt)$ and then integrating over $\mathcal{T}$. We obtain

$$N(M) = \int_{\mathcal{T}} F_r(t) \cdot H_r(t)^2 E(-Mt)\,dt\,, \tag{1.17}$$

where dt is the Haar measure on $\mathcal{T}$ of total mass one. If we are able to estimate $F_r(t)$ by simpler functions with a decently small error term, then we can use (1.17) to derive an asymptotic formula for $N(M)$ as $r \to \infty$.

The circle $\mathbb{T}$ is the Pontryagin dual of $\mathbb{Z}$. One natural candidate for $\mathcal{T}$, therefore, is the Pontryagin dual of $\mathbf{A}$ as a discrete additive group. Because $\mathbf{A}$ is an $\mathbb{F}_p$-vector space, the Pontryagin dual of $\mathbf{A}$ is naturally identified with the linear dual $\mathbf{A}^* = \mathrm{Hom}_{\mathbb{F}_p}(\mathbf{A}, \mathbb{F}_p)$. One can realize $\mathbf{A}^*$ by constructs which exhibit it as an analogue for the unit interval. Let $k = \mathbb{F}_q(T)$ be the field of fractions of $\mathbf{A}$, and let '∞' denote the infinite place of k relative to T (see §2.1). The completion k_∞ is the polynomial analogue of the real numbers; and (cf. Proposition 2.2) $\mathbf{A}^*$ is isomorphic as a topological $\mathbf{A}$-module to the maximal ideal $\mathcal{P}_\infty = \{t \in k_\infty : |t|_\infty < 1\}$ of the ring of integers in k_∞.

One can use $\mathcal{T} = \mathcal{P}_\infty$ to develop asymptotic formulas for the number of representations by sums of d-th powers in $\mathbf{A}$ (Car 1971a, Webb 1973a, Kubota 1974), for the number of 3-primes polynomials of given degree in $\mathbf{A}$ (Hayes 1966, Car 1971b), and for the number of representations by sums of d-th powers of irreducibles in $\mathbf{A}$ (Webb 1973b, Car 1972, 1981, Pfander 1981). Nevertheless, as we hope to convince the reader, there is a much more natural choice for $\mathcal{T}$. In this book, the circle method is based on the *adèle class group* $\mathbb{A}_k/k$ of the rational function field k, which is a compact k-vector space. As we show in Chapter 4, $\mathbb{A}_k/k$ is isomorphic to the Pontryagin dual of k as a discrete additive group.

After Weil's pioneering use of adèlic analysis to give new proofs of Siegel's theorems on quadratic forms (see Weil 1962, 1965), the possibility of formulating the circle method in adèlic terms occurred to many people. This possibility was actually carried out in detail for the classical Waring problem by Mars (1973) (see also Hayes 1972). The unit circle $\mathbb{T}$ is replaced by the adèle class group $\mathbb{A}_{\mathbb{Q}}/\mathbb{Q}$ of $\mathbb{Q}$, which is a compact $\mathbb{Q}$-vector space. This adèlic approach has many technical and conceptual advantages over the classical circle method, but it does not lead to new theorems. However, it may well provide a framework for new results in the future. In the next section, we will describe some of the advantages of using A_k/k instead of $\mathcal{P}_\infty$ for the polynomial 3-primes problem.

Exercises 1.4

1. Use the fact that a linear form on a subspace of $\mathbf{A}$ can be extended to all of $\mathbf{A}$ to show that $\mathbf{A}^*$ is a divisible $\mathbf{A}$-module.

1.5 The singular series

After the choice of a generator T of k, there is a canonical character $E : \mathsf{A}_k \to \mathbf{T}$ with $k \subseteq \ker(E)$ (see §4.3). In Chapter 7, we apply (1.17) with this character and $\mathcal{T} = \mathsf{A}_k/k$ to prove an asymptotic formula for $N(M)$ with a very good error term:

$$N(M) = \frac{1}{r} L_q(r-1)^2 S(M) + O\left(\frac{q^{7r/4}}{(q-1)(r-1)}\right) \tag{1.18}$$

where

$$L_q(r) = \sum_{i=1}^{r} \frac{q^i}{i} \tag{1.19}$$

and where $S(M)$ is the 'singular series'

$$S(M) = \sum_{A/Q} \frac{\mu(Q)}{\Phi(Q)^3} \cdot \widetilde{E}(-AM/Q)\,. \tag{1.20}$$

The character $\widetilde{E}$ is the restriction of E to the 'finite adèles' (see §4.1). The summation in (1.20) is over all reduced fractions A/Q with $A, Q \in \mathbf{A}$, Q positive, and $\deg A < \deg Q$; and the functions $\mu(Q)$ and $\Phi(Q)$ are the polynomial Möbius and Euler functions respectively. As we prove in §4.4, the singular series is absolutely convergent.

Some readers may not comprehend the full intentional halo surrounding the word 'singular'. For their benefit, we quote from *The Red Headed League* by A. Conan Doyle:

I know, my dear Watson, that you share my love of all that is bizarre and outside the conventions and humdrum routine of everyday life. You have shown your relish for it by the enthusiasm which has prompted you to chronicle, and, if you will excuse my saying so, somewhat to embellish so many of my own little adventures ... Now, Mr. Jebez Wilson here has been good enough to call upon me this morning, and to begin a narrative which promises to be one of the most SINGULAR which I have listened to for some time.

According to Vaughan (1981), the genesis of the name 'singular series' was not the unusual form of the classical analogues of (1.20) but rather the fact that they were derived from the singular points of an analytic function. Even so, in the 1920s, these were not the sort of series that the analyst routinely wrote down.

Let us take a careful look at the series (1.20). It appears to be the Fourier series of a function on some compact, abelian group G. Because (1.20) is absolutely convergent, that function is continuous on G and must be the extension to G of a function $S(M)$ on positive polynomials. The group G must therefore contain $\mathbf{A}$, presumably as a discrete subgroup since there is no other natural topology on $\mathbf{A}$. Since the singular series $S(M)$ extends to a continuous function on a compactification of $\mathbf{A}$, it is *almost periodic* as a function on positive polynomials M (see Loomis 1953, §41). The reader who wishes to explore the fascinating connections between analytic additive number theory and the theory of almost periodic functions should consult Mackey (1978).

Which group is G? The fractions A/Q clearly run over the discrete group $k/\mathbf{A}$, so G must be the Pontryagin dual of $k/\mathbf{A}$. Let $\mathbf{M}_{\text{fin}}$ be the set of positive irreducibles in $\mathbf{A}$; and for $P \in \mathbf{M}_{\text{fin}}$, let $\mathcal{O}_P$ be the ring of integers in the completion k_P of k at P. Each $\mathcal{O}_P$ is a compactification of $\mathbf{A}$, and so $\mathbf{A}$ injects as a discrete group along the diagonal of the product

$$\mathcal{O} = \prod_{P \in \mathbf{M}_{\text{fin}}} \mathcal{O}_P .$$

The compact ring $\mathcal{O}$ is just the closure of $\mathbf{A}$ in the ring of finite adèles. One can show that $k/\mathbf{A}$ is the Pontryagin dual of $\mathcal{O}$, which implies that $\mathcal{O}$ is isomorphic to G (see §4.4).

Having identified G in explicit form, it is natural to try to recognize the function defined on $\mathcal{O}$ by the series (1.20). Can we construct it in some intrinsic way without reference to the series (1.20)? For every $P \in \mathbf{M}_{\text{fin}}$, let W_P be the closure in $\mathcal{O}_P$ of the set of positive irreducibles in $\mathbf{A}$, and let

$$V_P = W_P \times W_P \times W_P .$$

Let $V \subset \mathcal{O}^3$ be the product of the sets V_P over all $P \in \mathbf{M}_{\text{fin}}$, and let h be the restriction to V of the function which maps $\tilde{\mathbf{t}} = (\tilde{t}_1, \tilde{t}_2, \tilde{t}_3) \in \mathcal{O}^3$ to $\tilde{t}_1 + \tilde{t}_2 + \tilde{t}_3 \in \mathcal{O}$. As we show in §4.4, the function h has a Radon–Nikodym derivative $\theta(\tilde{t})$ in Haar measure that effects the change of variables

$$\int_V f(\tilde{t}_1 + \tilde{t}_2 + \tilde{t}_3)\, d\rho_V(\tilde{\mathbf{t}}) = \int_{\mathcal{O}} f(\tilde{t})\theta(\tilde{t})\, d\tilde{t} \tag{1.21}$$

for every $f \in L^1(\mathcal{O})$. The measure $d\tilde{t}$ on $\mathcal{O}$ is Haar measure normalized so that $\mathcal{O}$ has unit total mass. The measure ρ_V is the product over $P \in$

$\mathbf{M}_{\text{fin}}$ of the Haar measure on $\mathcal{O}_P^3$ restricted to V_P and normalized so that each V_P has unit total mass. We can actually compute $\theta(\tilde{t})$ explicitly (see Theorem 4.15). For $\tilde{t} = (t_P)_{P \in \mathbf{M}_{\text{fin}}}$,

$$\theta(\tilde{t}) = \prod_{\substack{P \\ t_P \in \mathbf{U}_P}} \left(1 + \frac{1}{(|P|-1)^3}\right) \prod_{\substack{P \\ t_P \in P\mathcal{O}_P}} \left(1 - \frac{1}{(|P|-1)^2}\right)$$

the products extending over all $P \in \mathbf{M}_{\text{fin}}$. We can also compute the Fourier series of $\theta(\tilde{t})$. At $\tilde{t} = M$, it is precisely the singular series $S(M)$.

One can also prove an asymptotic formula for the number of representations of a polynomial $M \in \mathbf{A}$ as a sum of s d-th powers of polynomials of degree $\{\deg M/d\}$ or less (cf. Chapter 8). That asymptotic formula also involves a singular series, and once again we can interpret it as a Fourier series on the ring $\mathcal{O}$. It is, in fact, the Fourier series of the Radon–Nikodym derivative in Haar measure of the function $(\tilde{t}_1, \tilde{t}_2, \ldots, \tilde{t}_s) \mapsto \tilde{t}_1^d + \tilde{t}_2^d + \cdots + \tilde{t}_s^d$ from $\mathcal{O}^s$ to $\mathcal{O}$. A singular series for strict representations by sums of squares in $\mathbf{A}_q$ was introduced by Carlitz in 1947 (Carlitz 1947a).

Hardy and Littlewood were certainly aware that the singular series which they encountered in their work on the classical Waring and Goldbach problems were products over all prime numbers p of p-adic density functions. Such densities are also central to the work of Siegel on quadratic forms. In fact, the famous Siegel mass formula shows how to average the number of representations of a positive integer n over a given genus of positive definite quadratic forms in s variables so that this average divided by $n^{(s/2)-1}$ becomes almost periodic in n and therefore actually equal to the sum of the corresponding singular series!

Change of variable formulas like (1.21) are at the centre of our treatment of local and global singular series. We have been careful to present the local theory in Chapters 2 and 3 so that it is valid over any compact local ring as a base. Therefore, the computations of local singular series in §§2.4–2.5 and in §§3.3–3.4 apply equally well to the classical Waring and 3-primes problems and their polynomial analogues. Since it requires little additional effort to do so, the Waring problem computations in §3.3 are actually carried out for the diagonal polynomial

$$\gamma_1 X_1^{d_1} + \gamma_2 X_2^{d_2} + \cdots + \gamma_s X_s^{d_s},$$

the coefficients $\gamma_1, \gamma_2, \ldots, \gamma_s$ belonging to the group of units of the local ring. We specialize to the case $d = d_1 = d_2 = \cdots = d_s$ in §3.4.

Let $D \subset \mathbb{A}_k$ be a fundamental domain for the additive action of k. The functions $F_r(t)$ and $H_r(t)$ of §1.4 above make sense for any $t \in \mathbb{A}_k$; and (1.17) with $\mathcal{T} = \mathbb{A}_k/k$ becomes

$$N(M) = \int_D F_r(t) \cdot H_r(t)^2 E(-Mt)\, dt\,. \tag{1.22}$$

In Chapter 6, we use the Weil Riemann hypothesis to estimate $F_r(t)$. The resulting approximation is good when the denominator

$$\tilde{\mathfrak{n}}(t) = \prod_{P \in \mathbf{M}_{\mathrm{fin}}} P^{\max\{0, -\mathrm{v}_P(t_P)\}}$$

of the adèle t satisfies

$$\deg \tilde{\mathfrak{n}}(t) \leq r/2 \qquad \text{and} \qquad \mathrm{v}_\infty(t_\infty) > \deg \tilde{\mathfrak{n}}(t) + r/2,$$

where v_P and v_∞ are the normalized valuations at P and at ∞ respectively. The subset D of all $t \in \mathbb{A}_k$ which satisfy these two relations is the analogue of the Farey dissection of the unit interval, and this D is indeed a fundamental domain for the action of k. One of the advantages of the choice $\mathcal{T} = \mathbb{A}_k/k$ is that we are able to replace the finite part of D in (1.22) by the whole ring of finite adèles without degrading the error term. The asymptotic formula (1.18) then follows by Fourier inversion via the change of variable (1.21). See Chapter 7 for the details.

As we show in §5.4, the singular series $S(M)$ of (1.18) as a function on *odd* polynomials M is bounded from below by a non-zero positive constant which is independent of q. The implied constant in the error term is also independent of q. In order to deduce Theorem 1.22 from (1.18) therefore, it remains only to find good bounds for the function $L_q(r)$.

Lemma 1.24. *For all integers $r > 2$,*

$$\left(\frac{q}{q-1}\right)\frac{q^r}{r} \leq L_q(r) \leq \frac{3}{2}\left(\frac{q}{q-1}\right)\frac{q^r}{r}. \tag{1.23}$$

Proof. The inequality is true for $r = 2$ by direct substitution. Proceeding by induction, it suffices to show that for $r \geq 2$,

$$\left(\frac{q}{q-1}\right)\frac{q^{r+1}}{r+1} \leq \left(\frac{q}{q-1}\right)\frac{q^r}{r} + \frac{q^{r+1}}{r+1} \tag{1.24}$$

and

$$\frac{3}{2}\left(\frac{q}{q-1}\right)\frac{q^r}{r} + \frac{q^{r+1}}{r+1} \leq \frac{3}{2}\left(\frac{q}{q-1}\right)\frac{q^{r+1}}{r+1}. \tag{1.25}$$

Since the right hand side of (1.24) equals

$$\left(\frac{q}{q-1}\right)\frac{q^{r+1}}{r+1}\left(\frac{rq+1}{rq}\right),$$

that inequality is clearly valid for all $r \geq 2$. Now the left hand side of (1.25) equals

$$\frac{3}{2}\left(\frac{q}{q-1}\right)\frac{q^{r+1}}{r+1}\left(1-\frac{r(q-1)-3}{3rq}\right),$$

which implies the validity of (1.25) whenever $r(q-1) \geq 3$. This last inequality fails only for $r = q = 2$. But $L_2(3) = 20/3 < 8$, the value of the right hand side of (1.23) at $r = q = 2$. ∎

Exercises 1.5

1. Show that $W_P = \{P\} \cup \mathbf{U}_P$ where $\mathbf{U}_P$ is the group of units of the ring $\mathcal{O}_P$. Let

$$W \stackrel{\text{def}}{=} \prod_{P \in \mathbf{M}_{\text{fin}}} W_P \quad \text{and} \quad \mathbf{U} \stackrel{\text{def}}{=} \prod_{P \in \mathbf{M}_{\text{fin}}} \mathbf{U}_P .$$

Show that W differs from $\mathbf{U}$ by a set of measure zero in W. Show further that W is the closure in $\mathcal{O}$ of the set of positive irreducibles in $\mathbf{A}$. [Hint: Use Theorem 5.8.]

2

Local singular series

2.1 Fundamentals

Let $\mathbb{F}_q$ be the finite field of q elements, let T be transcendental over $\mathbb{F}_q$, and let k be the rational function field $\mathbb{F}_q(T)$. We assume that the reader is familiar with the concepts of discrete valuations of a field and the associated 'places' of the field (cf. Borevich and Shafarevich 1966, Serre 1979). With one exception, the places of k correspond one-to-one with the non-zero prime ideals of the polynomial ring $\mathbf{A} = \mathbb{F}_q[T]$. The unique exception is the 'place at infinity' which we denote by the symbol '∞'. The valuation v_∞ corresponding to ∞ is defined as follows: If $X = R/S$ with $R, S \in \mathbf{A}$, $R \neq 0$, $S \neq 0$, then

$$\mathrm{v}_\infty(X) = -\deg X$$

where

$$\deg X = (\deg R) - (\deg S). \tag{2.1}$$

The *absolute value* of X at ∞ is defined by

$$|X| = q^{\deg X}. \tag{2.2}$$

Let k_∞ denote the completion of k at ∞. Then every element $X \in k_\infty^\times$ may be written uniquely as a Laurent series

$$X = \sum_{\nu=n}^{\infty} \alpha_\nu \left(\frac{1}{T}\right)^\nu \tag{2.3}$$

in $1/T$, where $n = \mathrm{v}_\infty(X)$ and $\alpha_\nu \in \mathbb{F}_q$ for all $\nu \geq n$. For us, k_∞ is the analogue of the field of real numbers, and (2.3) is the analogue of the infinite decimal expansion of a real number. Following Artin (1924), we introduce a function sgn: $k_\infty \longrightarrow \mathbb{F}_q$ which is analogous to the usual sign function on $\mathbb{R}$. We put $\text{sgn}(0) = 0$ and for $X \in k_\infty^\times$, we define

$$\text{sgn}(X) = \alpha_n ,$$

the leading coefficient of the Laurent series (2.3). We say that X is *positive* if $\text{sgn}(X) = 1$. In particular, a polynomial in T (i.e. an element of $\mathbf{A}$) is positive if and only if it is monic.

Let $\mathbf{M}$ denote the set of places of k, and let $\mathbf{M}_{\text{fin}} = \mathbf{M} - \infty$. A place in $\mathbf{M}_{\text{fin}}$ will be called *finite.* Such a place determines a unique non-zero prime ideal in the polynomial ring $\mathbf{A}$ and therefore a unique positive irreducible $P \in \mathbf{A}$. Although it is an abuse of language, we will use the letter 'P' to denote both a positive irreducible and the place associated with the ideal $P\mathbf{A}$. Thus if v_P is the normalized valuation at the place P and $X \in k^\times$, then $\text{v}_P(X)$ is the exponent of the power of P in X.

Let k_P be the completion of k at the finite place P, let $\mathcal{O}_P$ be the valuation ring of k_P (i.e. $\mathcal{O}_P = \{X \in k_P : \text{v}_P(X) \geq 0\}$), and let $\mathbf{K}(P) = \mathcal{O}_P/P\mathcal{O}_P$ be the residue class field at P. Let $\text{can}_P : \mathcal{O}_P \longrightarrow \mathbf{K}(P)$ be the canonical map. The completion k_P is analogous to the field $\mathbb{Q}_p$ of p-adic numbers. The analogue for $X \in k_P^\times$ of the usual p-adic expansion is

$$X = \sum_{\nu=n}^{\infty} C_\nu P^\nu \tag{2.4}$$

where $n = \text{v}_P(X)$ and where $C_\nu \in \mathbf{A}$ satisfies $\deg C_\nu < \deg P$ for all $\nu \geq n$. We may think of the coefficients C_ν as being representatives of elements of the finite field $\mathbf{A}/P\mathbf{A} \simeq \mathbf{K}(P)$. We also have another type of Laurent expansion which identifies k_P as the ring of Laurent series over $\mathbf{K}(P)$. Since k_P has positive characteristic, the residue class field $\mathbf{K}(P)$ is naturally embedded as a subfield of k_P (cf. Serre 1979, II, §4). Let α be the root of unity in k_P such that $\text{can}_P(\alpha) = \text{can}_P(T)$. Then α is a root of P, and every $X \in k_P^\times$ has an expansion

$$X = \sum_{\nu=n}^{\infty} \gamma_\nu (T - \alpha)^\nu$$

where $n = \text{v}_P(X)$ and $\gamma_\nu \in \mathbf{K}(P)$ for $\nu \geq n$. Since $q_P = q^{\deg P}$ is the number of elements in the residue class field $\mathbf{K}(P)$,

$$|X|_P = q_P^{-n} = q_P^{-\text{v}_P(X)}$$

is the *normalized absolute value* of $X \in k_P$.

The reader should note that many of the above definitions depend on the choice of the generator T for the rational function field k.

In the rest of this chapter and also in Chapter 3, we will introduce some basic harmonic analysis for k_∞ and the fields k_P. The use of harmonic analysis on local fields to study problems in analytic number theory originates with Tate's thesis (Tate 1950). The results we will state are valid, in fact, for any non-archimedean local field, and it is instructive to introduce these results in the general context. Therefore, the following notation will be in force until the end of Chapter 3: let k_* denote a complete valued field with discrete valuation v_*, valuation ring $\mathcal{O}_*$, uniformizing parameter π_*, and *finite* residue class field $\mathbf{K}(*) = \mathcal{O}_*/\pi_*\mathcal{O}_*$. Let q_* be the order of $\mathbf{K}(*)$, and define the normalized absolute value of $X \in k_*$ by

$$|X|_* = q_*^{-\mathrm{v}_*(X)}.$$

The distance function δ defined for $X,Y \in k_*$ by

$$\delta(X,Y) = |X - Y|_* \tag{2.5}$$

provides a metric topology on k_* which is *locally compact.* In fact, $\mathcal{O}_*$ is compact in this topology, and thus k_* is naturally a topological field with a locally compact topology.

Exercises 2.1

1. Let $X \in k_\infty^*$, and let r be a positive integer prime to the characteristic of k. Show (using Newton's method or Hensel's lemma) that X has an r-th root in $k_\infty^* \iff \mathrm{v}_\infty(X) \in n\mathbb{Z}$ and $\mathrm{sgn}(X)$ is an n-th power in $\mathbb{F}_q^\times$.
2. Fill in the details for the following proof that $\mathcal{O}_*$ is compact. First, $\mathcal{O}_*/\pi_*^\nu\mathcal{O}_*$ is finite for all $\nu \geq 0$. Second, the cosets of $\pi^\nu\mathcal{O}_*$ provide an open covering of $\mathcal{O}_*$ by sets of diameter $q_*^{-\nu}$. Thus $\mathcal{O}_*$ is a totally bounded metric space.

2.2 Characters

Let G be a locally compact abelian group. The *character group* $\widehat{G}$ of G is the set of all continuous group homomorphisms from G to the circle group $\mathbb{T} = \{z \in \mathbb{C} : |z| = 1\}$. One knows (Loomis 1953, VII) that if $\widehat{G}$ is appropriately topologized, then $\widehat{G}$ itself becomes a locally compact group; and one knows further that the character group of $\widehat{G}$ with this topology is naturally isomorphic to G itself. We wish to study the characters of the additive group k_*^+ and the multiplicative group $k_*^\times$ of our local field k_*. Most of the time, 'Λ' will denote a character of k_*^+ and 'χ' will denote a character of $k_*^\times$. We let 'Λ_0' (resp. 'χ_0') denote the principal (i.e. trivial) character of k_*^+ (resp. $k_*^\times$).

Let Λ be any additive character. Consider the chain of fractional ideals:

$$\ldots \supseteq \pi_*^{-2}\mathcal{O}_* \supseteq \pi_*^{-1}\mathcal{O}_* \supseteq \mathcal{O}_* \supseteq \pi_*\mathcal{O}_* \supseteq \pi_*^2\mathcal{O}_* \supseteq \ldots .$$

Since Λ is continuous, there exists a ν such that Λ is trivial on $\pi_*^\nu\mathcal{O}_*$. Let

$$n(\Lambda) \stackrel{\text{def}}{=} \min\{\nu : \Lambda(x) = 1 \text{ for every } x \in \pi_*^\nu\mathcal{O}_*\}. \tag{2.6}$$

We call $n(\Lambda)$ the *conductor* of Λ. (We note that frequently in the literature the 'conductor' is the ideal $\pi_*^\nu\mathcal{O}_*$ rather than the integer ν, but the latter is more convenient for us.)

Similarly, let χ be a multiplicative character. We define

$$\mathbf{U}_* = \{x \in k_*^\times : \mathrm{v}(x) = 0\} = \mathcal{O}_* - \pi_*\mathcal{O}_*. \tag{2.7}$$

Let $\mathbf{U}_*^{(0)} = \mathbf{U}_*$, and for $\nu \geq 1$ define

$$\mathbf{U}_*^{(\nu)} = 1 + \pi_*^\nu\mathcal{O}_*. \tag{2.8}$$

This gives a filtration of multiplicative subgroups of $k_*^\times$:

$$\mathbf{U}_* = \mathbf{U}_*^{(0)} \supseteq \mathbf{U}_*^{(1)} \supseteq \mathbf{U}_*^{(2)} \supseteq \ldots , \tag{2.9}$$

and we define

$$m(\chi) \stackrel{\text{def}}{=} \min\{\nu : \chi(x) = 1 \text{ for every } x \in \mathbf{U}_*^{(\nu)}\}. \tag{2.10}$$

Again, $m(\chi)$ exists by the continuity of χ, and it is called the *conductor* of χ. We observe that $-\infty \leq n(\Lambda) < \infty$ whereas $0 \leq m(\chi) < \infty$.

Given an additive character Λ and an element $a \in k_*^\times$, let Λ^a be the character defined by

$$\Lambda^a(x) \stackrel{\text{def}}{=} \Lambda(ax) \tag{2.11}$$

for all $x \in k_*$. Since $\Lambda^a(\pi_*^\nu\mathcal{O}_*) = 1$ if and only if $\Lambda(\pi_*^{\nu+\mathrm{v}_*(a)}\mathcal{O}_*) = 1$, we have

$$n(\Lambda^a) = n(\Lambda) - \mathrm{v}_*(a). \tag{2.12}$$

Thus, if $\hat{k}_*^+$ is non-trivial (as indeed it is!), then there are additive characters with any given conductor.

In §2.3, we will need to know the relationship between the groups k_*^+ and $\hat{k}_*^+$. To this end, let Λ be any non-trivial character of k_*^+, and define a mapping $\phi_\Lambda : k_*^+ \to \hat{k}_*^+$ as follows: for any $a \in k_*$, $\phi_\Lambda(a) = \Lambda^a$. The result is then:

Proposition 2.1. *The mapping $\phi_\Lambda : k_*^+ \to \hat{k}_*^+$ is an algebraic and topological isomorphism.*

Proof. This is Theorem 7-1-10 in Goldstein (1971). We note here that the only difficulty is in showing that ϕ_Λ is onto, which requires establishing that $\phi_\Lambda(k_*)$ is both closed and dense in $\hat{k}_*^+$. ∎

Finally, we wish to use the isomorphism ϕ_Λ to introduce another isomorphism which we will need in §2.3. In general, if H is a subgroup of G, we define

$$H^\perp \stackrel{\text{def}}{=} \{f \in \widehat{G} : f(h) = 1 \text{ for all } h \in H\}.$$

We use this construct to prove the following.

Proposition 2.2. *The group $\widehat{\mathcal{O}}_*$ is isomorphic to $k_*^+/\mathcal{O}_*$, and hence is discrete.*

Proof. Let Λ be a non-trivial additive character with conductor $n(\Lambda) = 0$. Let $\psi_\Lambda : k_*^+ \to \hat{k}_*^+/\mathcal{O}_*^\perp$ be the composition of ϕ_Λ (as in Proposition 2.1) followed by the canonical surjection $\hat{k}_*^+ \to \hat{k}_*^+/\mathcal{O}_*^\perp$. The kernel of ψ_Λ is just $\{x \in k_*^+ : \Lambda(xy) = 1 \text{ for all } y \in \mathcal{O}_*\}$, and since $n(\Lambda) = 0$, this is just $\mathcal{O}_*$ itself. Finally since $\hat{k}_*^+/\mathcal{O}_*^\perp$ is isomorphic to $\widehat{\mathcal{O}}_*$ (Loomis 1953, §35), we have established a continuous group isomorphism from $k_*^+/\mathcal{O}_*$ onto $\widehat{\mathcal{O}}_*$. Now since $\mathcal{O}_*$ is both open and closed in k_*^+, our group isomorphism is bi-continuous. ∎

Exercises 2.2

1. Let G be a locally compact group. The topology on $\widehat{G}$ is defined as follows: if N is a neighbourhood of $1 \in \mathbb{T}$ and K is a compact subset of G, then a basic open neighbourhood of $\alpha \in \widehat{G}$ is $\{\beta \in \widehat{G} : \beta(x)\alpha(x)^{-1} \in N \text{ for all } x \in K\}$. This is called the 'topology of uniform convergence on compact sets'. Show that if G is compact, then $\widehat{G}$ is discrete.

2.3 Haar measure

On any locally compact abelian group G, one can define an invariant measure dt which is unique up to multiplication by a strictly positive constant (Nachbin 1965). The measure dt is called a *Haar measure* on G, and the invariance of the measure is equivalent to the identity

$$\int_G f(at)\,dt = \int_G f(t)\,dt$$

for all $f \in L^1(dt)$ and all $a \in G$. A Haar measure always assigns a finite measure to any compact subset of G. In particular when G itself is compact, the measure dt is finite, and one may then normalize dt so that the total mass of G is one.

For a given $f \in L^1(dt)$, the *Fourier transform* $\hat{f}$ of f is a function on the character group defined for $\chi \in \widehat{G}$ by

$$\hat{f}(\chi) \stackrel{\text{def}}{=} \int_G f(t)\chi(t)\,dt\,.$$

Let $d\chi$ be a Haar measure on $\widehat{G}$. The Fourier transform $\hat{f}$ is always a continuous function on $\widehat{G}$, but $\hat{f}$ need not be integrable with respect to the measure $d\chi$. However, if $\hat{f} \in L^1(d\chi)$, then one can recover f from $\hat{f}$ by the *Fourier inversion theorem*: there is a constant c which depends on the choice of the Haar measures dt and $d\chi$ but not on f such that for almost all $t \in G$

$$f(t) = c \int_{\widehat{G}} \hat{f}(\chi)\overline{\chi(t)}\,d\chi,$$

where the bar over the $\chi(t)$ indicates complex conjugation (see Loomis 1953). When dt and $d\chi$ are chosen so that $c = 1$, these two measures are called *dual measures*. One can show that when G is compact and dt has total mass one, then dt is dual to the counting measure on the discrete group $\widehat{G}$.

We denote the Haar measure of the additive group k_*^+ by d_*t, and we assume this measure normalized so that the compact open subgroup $\mathcal{O}_*$ has measure one. The defining properties of the measure d_*t become

$$\int_{\mathcal{O}_*} d_*t = 1$$

and

$$\int f(t+a)\,d_*t = \int f(t)\,d_*t \tag{2.13}$$

for all $f \in L^1(d_*t)$ and all $a \in k_*^+$. Using (2.13) and counting cosets of $\pi_*^\nu\mathcal{O}_*$ in $\mathcal{O}_*$ for $\nu \geq 0$ and cosets of $\mathcal{O}_*$ in $\pi_*^\nu\mathcal{O}_*$ for $\nu < 0$, one sees easily that

$$meas(\pi_*^\nu\mathcal{O}_*) = q_*^{-\nu} \tag{2.14}$$

for all $\nu \in \mathbb{Z}$, or more generally that

$$meas(c\,\mathcal{O}_*) = |c|_* \tag{2.15}$$

for any $c \in k_*^\times$. Now (2.15) and the uniqueness of d_*t enable us to prove

$$\int f(ct)\,d_*t = \frac{1}{|c|_*}\int f(t)\,d_*t \tag{2.16}$$

for all $f \in L^1(d_*t)$ and all $c \in k_*^\times$.

We will now state the Fourier inversion theorem for k_*^+ in a form which is very easy to use. By Proposition 2.1, we know that a non-trivial character Λ of k_*^+ determines an isomorphism of $\hat{k}_*^+$ with k_*^+ itself. Via this isomorphism, the Fourier transform $\hat{f}$ of a function $f \in L^1(d_*t)$ takes the form

$$\hat{f}(x) = \int f(t)\Lambda(xt)\, d_*t$$

for all $x \in k_*^+$. If $\hat{f} \in L^1(d_*t)$ then *its* Fourier transform via the conjugate of Λ takes the form

$$\hat{\hat{f}}(t) = \int \hat{f}(x)\Lambda(-xt)\, d_*x$$

for all $t \in k_*^+$.

Proposition 2.3. (Fourier Inversion) *If $\hat{f} \in L^1(d_*x)$ for continuous $f \in L^1(d_*t)$, then*

$$\hat{\hat{f}}(t) = q_*^{-n(\Lambda)} f(t) \tag{2.17}$$

for all $t \in k_*^+$, where $n(\Lambda)$ is the conductor of Λ as defined in (2.6).

Proof. From the general theory outlined at the beginning of this section, we know that $\hat{\hat{f}} = c_\Lambda f$ for some constant c_Λ which is independent of f. Taking f to be the characteristic function of $\mathcal{O}_*$, one easily computes that $\hat{f}$ is the characteristic function of $\pi_*^{n(\Lambda)}\mathcal{O}_*$. From this it follows that

$$\hat{\hat{f}}(t) = meas(\pi_*^{n(\Lambda)}\mathcal{O}_*) \cdot f(t) = q_*^{-n(\Lambda)} f(t)$$

as desired. ■

In real analysis, one sometimes deduces a Fourier series expansion theorem for periodic functions on $\mathbf{R}$ from the Fourier inversion theorem for functions in $L^1(\mathbf{R})$. We will now follow the same sort of procedure in using Proposition 2.3 to prove a series expansion theorem for continuous functions on the compact group $\mathcal{O}_*$.

For each $\nu \geq 0$, choose units $a \in \mathbf{U}_*$ so that a/π^ν runs through a set of representations for those cosets of $\pi_*^{-\nu}\mathcal{O}_*/\mathcal{O}_*$ which contain elements of absolute value q_*^ν. Such cosets are called *primitive.* The union of these sets for all $\nu \geq 0$ constitutes a set of representatives for the cosets $k_*^+/\mathcal{O}_*$. In what follows, a summation over elements a/π^ν is understood to be a summation over such a set of representatives for $k_*/\mathcal{O}_*$.

Let $\Lambda \in \hat{k}_*^+$ have conductor $n(\Lambda) = 0$. The isomorphism $k_*^+/\mathcal{O}_* \to \widehat{\mathcal{O}}_*$ of Proposition 2.2 can be made explicit in the following way: the coset

$c = a/\pi_*^\nu + \mathcal{O}_*$, with $\nu \geq 0$ and $a \in \mathbf{U}_*$, induces the character of $\mathcal{O}_*$ given by

$$t \longmapsto \Lambda(at/\pi_*^\nu)$$

for $t \in \mathcal{O}_*$. We can therefore write the Fourier coefficient of a continuous function f on $\mathcal{O}_*$ as

$$c_f(a/\pi_*^\nu) = \int_{\mathcal{O}_*} f(t)\Lambda(at/\pi_*^\nu)\, d_*t \tag{2.18}$$

where a/π_*^ν is one of our standard representatives for $k_*^+/\mathcal{O}_*$.

Proposition 2.4. *Let $n(\Lambda) = 0$, and suppose that f is a continuous $\mathbb{C}$-valued function on $\mathcal{O}_*$ such that*

$$\sum_{a/\pi_*^\nu} |c_f(a/\pi_*^\nu)| < \infty.$$

Then

$$f(t) = \sum_{a/\pi_*^\nu} c_f(a/\pi_*^\nu) \cdot \Lambda(-at/\pi_*^\nu) \tag{2.19}$$

for all $t \in \mathcal{O}_*$.

Proof. Extend f to all of k_* by defining $f(t) = 0$ for $t \in k_* - \mathcal{O}_*$. Since $\mathcal{O}_*$ is open in k_*, f is continuous on k_* and

$$\hat{f}(x) = \int_{\mathcal{O}_*} f(t)\Lambda(tx)\, d_*t.$$

Since $n(\Lambda) = 0$, we see that for $x = b + y$ with $y \in \mathcal{O}_*$

$$\hat{f}(b+y) = \int_{\mathcal{O}_*} f(t)\Lambda(t(b+y))\, d_*t = \int_{\mathcal{O}_*} f(t)\Lambda(tb)\, d_*t = \hat{f}(b).$$

Hence $\hat{f}$ is periodic on k_* with $\mathcal{O}_*$ as a group of periods. Therefore, we have

$$\begin{aligned} \int |\hat{f}(x)|\, d_*x &= \sum_{a/\pi_*^\nu} \int_{a/\pi_*^\nu + \mathcal{O}_*} |\hat{f}(x)|\, d_*x \\ &= \sum_{a/\pi_*^\nu} |\hat{f}(a/\pi_*^\nu)| \cdot meas(a/\pi_*^\nu + \mathcal{O}_*) \end{aligned}$$

$$= \sum_{a/\pi_*^\nu} |c_f(a/\pi_*^\nu)| < \infty ,$$

since $meas(a/\pi_*^\nu + \mathcal{O}_*) = meas(\mathcal{O}_*) = 1$. Thus $\hat{f} \in L^1(d_*x)$, and so (2.17) applies with $n(\Lambda) = 0$, giving us

$$\begin{aligned} f(t) &= \int \hat{f}(x)\Lambda(-xt)\, d_*x \\ &= \sum_{a/\pi_*^\nu} \hat{f}(a/\pi_*^\nu) \int_{a/\pi_*^\nu + \mathcal{O}_*} \Lambda(-xt)\, d_*x \\ &= \sum_{a/\pi_*^\nu} \hat{f}(a/\pi_*^\nu)\Lambda(-at/\pi_*^\nu) \int_{\mathcal{O}_*} \Lambda(-xt)\, d_*x \\ &= \sum_{a/\pi_*^\nu} c_f(a/\pi_*^\nu)\Lambda(-at/\pi_*^\nu) \end{aligned}$$

because $\Lambda(-xt) \equiv 1$ for $x, t \in \mathcal{O}_*$. ■

The summation on the righthand side of (2.19) is called a local 'singular series'. This series is just the Fourier series of $f(t)$ as a function on the compact abelian group $\mathcal{O}_*$.

Exercises 2.3

1. Compute the Fourier coefficients $c(a/\pi_*^\nu)$ defined by (2.18) for the function $f(t) = |t|_*$ restricted to $\mathcal{O}_*$. [Hint: Establish the recurrence $c(a/\pi_*^{\nu+1}) = q^{-1}c(a/\pi_*^\nu)$.]

2.4 Local Radon–Nikodym derivatives

Given an integer $s \geq 1$, let V be a compact open subset of $\mathcal{O}_*^s$ and let $h : V \longrightarrow \mathcal{O}_*$ be a continuous map. Let ρ be the restriction of the product measure $\rho = d_*t_1 \times \cdots \times d_*t_s$ to V, normalized so that $\rho(V) = 1$. Then ρ projects under h to a measure $\mu(h, V)$ on $\mathcal{O}_*$. This projected measure is defined as follows: if f is any continuous $\mathbb{C}$-valued function on $\mathcal{O}_*$, then

$$\int_{\mathcal{O}_*} f\, d\mu(h, V) \stackrel{\text{def}}{=} \int_V f \circ h\, d\rho. \tag{2.20}$$

This definition extends in the familiar way from the space of all continuous functions on $\mathcal{O}_*$ to a natural space of integrable functions $L^1(\mu(h, V))$ on $\mathcal{O}_*$. It is clear that $\mu(h, V)$ is supported on the image $h(V)$ of the map h.

If $\mu(h, V)$ is absolutely continuous with respect to the measure d_*t (i.e. f is $\mu(h, V)$-negligible $\Rightarrow$ f is d_*t-negligible), then (cf. Loomis 1953, §15) there is a positive-valued function $\theta_*(t; h, V) \in L^1(d_*t)$ such that $\mu(h, V) = \theta_*(t; h, V)\, d_*t$. This function is called the *Radon–Nikodym derivative* of h. When it exists, we can rewrite (2.20) as a change of variables:

$$\int_V f(h(t_1, \ldots, t_s)) \frac{d_*t_1 \cdots d_*t_s}{meas(V)} = \int_{\mathcal{O}_*} f(t)\theta_*(t; h, V)\, d_*t$$

where $meas(V)$ is the measure of V in the product measure on $\mathcal{O}_*^s$. Intuitively, $\theta_*(t; h, V) = d\rho/d_*t$ is the amount of mass, relative to d_*t, which h projects onto small neighbourhoods of t.

We are interested primarily in the derivatives of polynomial functions in s variables with coefficients in $\mathcal{O}_*$. Such derivatives appear as local factors in the asymptotic formulas for the polynomial 3-primes and Waring problems. The derivatives of polynomial functions have some remarkable properties which we will now state (without proof) to help the reader gain insight into their behaviour. The proofs given in Hayes and Nutt (1982) over a p-adic field can be adapted to work over the general local field k_*. The reader should also consult §3 of Serre (1981).

Definition 2.5. *For $h \in \mathcal{O}_*[X_1, \ldots, X_s]$, the* singular set $S(h, V)$ *of h on V is defined by*

$$S(h, V) \stackrel{\text{def}}{=} \{P \in V : (\partial h/\partial t_i)(P) = 0 \text{ for all } 1 \le i \le s\}.$$

When $\text{char}(k_*) = 0$, Sard's theorem implies that $h(S(h, V)) \subset \mathcal{O}_*$ has d_*t-measure zero. However, if $\text{char}(k_*) = p > 0$, then it is possible that the singular set of h is all of $\mathcal{O}_*$! For $h \in \mathcal{O}_*[X_1, \ldots, X_s]$, assume that $h(S(h, V))$ has d_*t-measure zero. Then we have the following.

RD1 The derivative $\theta_*(t; h, V)$ exists.

RD2 The set $h(S(h, V))$ is *finite*. Further, $\theta_*(t; h, V)$ is *locally constant* on the complement of $h(S(h, V))$ in $\mathcal{O}_*$.

RD3 Let $t \in \mathcal{O}_* - h(S(h, V))$. Then $M_t = h^{-1}(t)$ is a $*$-adic analytic $(s-1)$-submanifold of $\mathcal{O}_*^s$ (cf. Serre 1965). Further, there is a natural $*$-adic analytic $(s-1)$-form ω on M_t such that

$$\theta_*(t; h, V) = \frac{1}{meas(V)} \int_{V \cap M_t} |\omega|_* .$$

The analytic $(s-1)$-form ω is defined as follows: for $P \in M_t$ choose i, $1 \le i \le s$, so that $(\partial h/\partial t_i)(P) \ne 0$. Then

$$\omega(P) = (-1)^i \cdot \frac{d_*t_1 \wedge \cdots \wedge d_*t_{i-1} \wedge d_*t_{i+1} \wedge \cdots \wedge d_*t_s}{\frac{\partial h}{\partial t_i}(P)}. \tag{2.21}$$

This definition does not depend on the choice of i.

Note that whereas $\theta_*(t; h, V)$ is initially defined only almost everywhere, Property **RD2** serves to fix its value at every point in the complement of the finite set $h(S(h, V))$.

We will not use these properties in subsequent chapters. In §3.3, we will use Proposition 2.4 to compute some specific examples of local derivatives, including the ones we need for the applications to the 3-primes and Waring problems. All these examples involve *diagonal* polynomials, i.e. polynomials of the form

$$h(X_1, X_2, \ldots, X_s) = \gamma_1 X_1^{d_1} + \gamma_2 X_2^{d_2} + \ldots + \gamma_s X_s^{d_s}$$

where $\gamma_i \in \mathbf{U}_*$ for $1 \le i \le s$. For diagonal polynomials, this Fourier series method is very powerful; and for large s, it reveals most of what we need to know about $\theta_*(t; h, V)$ (including its existence!). However, there is one further general property of derivatives of polynomial functions which we will need in our analysis of Waring's problem. This property is introduced with proof in the next section.

Exercises 2.4

1. Use the $*$-adic implicit function theorem (Serre 1965) to show that $M_{t,V} = V \cap M_t$ for a given $t \in h(V) - h(S(h, V))$ is an $(s-1)$-dimensional analytic submanifold of V.

2. Prove that (2.21) is independent of the choice of i.

3. Using Property **RD2** and (2.21), show for $t \in \mathcal{O}_* - h(S(h, V))$ that $t \in h(V) \iff \theta_*(t; h, V) \neq 0$.

4. Taking $s = 1$, let $h(X) = \gamma X^d$, where $\gamma \in \mathcal{O}_*$, $\gamma \neq 0$, and where $p \nmid d$ if k_* has characteristic $p > 0$. Show if $0 \neq t = \gamma x^d$, $x \in \mathcal{O}_*$, that

$$\theta_*(t; h, \mathcal{O}_*) = \#\mu_d(k_*) \cdot \frac{|t|_*^{\frac{1}{d}-1}}{|\gamma d|_*}$$

where $\mu_d(k_*)$ is the group of d-th roots of unity in k_*. [Hint: Use Property **RD3** for the case $s = 1$.] How does this formula compare with the change of variables $t = \gamma x^d$, $\gamma > 0$, for Riemann integration over the interval $[-1, 1]$?

5. Let k_*^{ac} be the algebraic closure of k_*. Let $h(X_1, \ldots, X_s)$ be a *separable* polynomial in s variables over k_*^{ac}. The singular set
$$S(h) = \{P \in (k_*^{ac})^s : (\partial h/\partial t_i)(P) = 0 \text{ for all } 1 \leq i \leq s\}$$
is an algebraic variety in the s-dimensional affine space over k_*^{ac}. Prove that the map $P \mapsto h(P)$ is constant on the components of $S(h)$. Conclude that $h(S(h))$ is finite. [Hint: Note that the differential $d_P h$ is zero at every point $P \in S(h)$.]

2.5 The case when $S(h, V)$ is empty

Throughout this section, we assume that the Radon–Nikodym derivative $\theta_*(t; h, V)$ of the polynomial $h \in \mathcal{O}_*[X_1, \ldots, X_s]$ viewed as a map from V to $\mathcal{O}_*$ exists. The theorem we wish to prove is most conveniently stated for unnormalized derivatives, and so we set $\theta_*^u(t; h, V) = meas(V) \cdot \theta_*(t; h, V)$.

Since V is compact and open, there is a non-negative integer ν_0 such that $P \in V \Rightarrow P + \pi_*^\nu \mathcal{O}_*^s \subset V$ for all $\nu \geq \nu_0$. A value for ν_0 is fixed for the remainder of this section.

For $\nu \geq 0$, let
$$\mathbf{red}_\nu : \mathcal{O}_*^s \longrightarrow \mathcal{O}_*^s / \pi_*^\nu \mathcal{O}_*^s = (\mathcal{O}_* / \pi_*^\nu \mathcal{O}_*)^s$$
be the reduction map modulo π_*^ν. When there is no danger of confusion, we write $\bar{P} = \mathbf{red}_\nu(P)$. Put $M_{t,V} = V \cap M_t$. For $\nu \geq \nu_0$, let $V(\nu) = \mathbf{red}_\nu(V)$; and for $t \in \mathcal{O}_*$, let
$$M_{t,V}(\nu) = \{Q \in V(\nu) : h(Q) \equiv t \pmod{\pi_*^\nu}\}.$$
Of course, $\mathbf{red}_\nu(M_{t,V}) \subseteq M_{t,V}(\nu)$, but there may well be points in $M_{t,V}(\nu)$ which do not come by reduction from $M_{t,V}$. We say that a point $Q \in M_{t,V}(\nu)$ is *liftable* if there is a point $P \in M_{t,V}$ such that $Q = \mathbf{red}_\nu(P)$.

The value $\theta_*^u(t; h, V)$ and the cardinalities $\#M_{t,V}(\nu)$ are intimately related. In fact,
$$\theta_*^u(t; h, V) = \lim_{\nu \to \infty} q_*^{(1-s)\nu} \cdot \#M_{t,V}(\nu) \tag{2.22}$$
for every value $t \in \mathcal{O}_*$ which is a point of continuity for $\theta_*^u(t; h, V)$. To verify (2.22), we note first that for $\nu \geq \nu_0$
$$h^{-1}(t + \pi_*^\nu \mathcal{O}_*) = \bigcup_P P + \pi_*^\nu \mathcal{O}_*^s \tag{2.23}$$
where P runs through any set of representative points in V for the cosets in $M_{t,V}(\nu)$. Since the union on the right in (2.23) is disjoint,
$$q_*^{-s\nu} \cdot \#M_{t,V}(\nu) \quad = \quad meas(h^{-1}(t + \pi_*^\nu \mathcal{O}_*))$$

$$= \int_{h^{-1}(t+\pi_*^\nu \mathcal{O}_*)} d_*z_1 \cdots d_*z_s \tag{2.24}$$
$$= \int_{t+\pi_*^\nu \mathcal{O}_*} \theta_*^u(t;h,V)\, d_*t\,.$$

Dividing both sides of this last equality by $q_*^{-\nu}$, which is the d_*t-measure of $t+\pi_*^\nu\mathcal{O}_*$, and taking the limit as $\nu \to \infty$, we obtain (2.22). If we believe Property **RD2** of §2.4, then we know that derivatives are locally constant, and hence continuous, almost everywhere. Therefore (2.22) and (2.25) show that knowing the function $\theta_*^u(t;h,V)$ is equivalent to knowing the counts $\#M_{t,V}(\nu)$ for all $t \in \mathcal{O}_*$ and all large values of ν. However, this fact is of little use for the purpose of computing $\theta_*^u(t;h,V)$ at a given $t \in \mathcal{O}_*$ unless one has some control over the sequence $\#M_{t,V}(\nu)$. Our aim in the remainder of this section is to show that when $S(h,V)$ is empty, the limit in (2.22) becomes stationary for large values of ν (cf. Shuck 1969).

Definition 2.6. *The function $F : \mathcal{O}_*^s \longrightarrow \mathbb{Z} \cup \{\infty\}$ defined by*

$$F(P) \stackrel{\text{def}}{=} \min_i \mathrm{v}_*\left(\frac{\partial h}{\partial x_i}(P)\right) \tag{2.25}$$

is continuous for the topology which $\mathbb{Z} \cup \{\infty\}$ inherits as a subset of the interval $[-\infty,+\infty]$. We put

$$\ell(h,V) \stackrel{\text{def}}{=} \sup_{P\in V} F(P)\,. \tag{2.26}$$

Lemma 2.7. *The value of $\ell(h,V)$ is a non-negative integer or $+\infty$. Further, $\ell(h,V) < +\infty$ if and only if $S(h,V)$ is empty.*

Proof. Since $F(P)$ is continuous and V is compact, there is a point $P_0 \in V$ where the supremum (2.26) is attained. Therefore $\ell(h,V) = F(P_0) \in \mathbb{Z} \cup \{+\infty\}$. Further, $F(P_0) = +\infty \iff P_0 \in S(h,V)$. ∎

Lemma 2.8. *Assume that $S(h,V)$ is empty. Assume further that $F(P)$ is constant on V, and let $\ell = \ell(h,V) < +\infty$ be its value. Then for any integer $\gamma \geq \max\{\ell+\nu_0, 2\ell+1\}$,*

$$\#M_{t,V}(\nu) = q_*^{(s-1)(\nu-\gamma)} \cdot \#M_{t,V}(\gamma) \tag{2.27}$$

for all $\nu \geq \gamma$.

Proof. For $\nu \geq \nu_0$, let $Y_{t,V}(\nu)$ be the reduction down to $V(\nu)$ of the points in $M_{t,V}(\nu+\ell)$. Given a point $\bar{P} \in Y_{t,V}(\nu)$, put $R = P + \pi_*^\nu U$ with $P \in V$ and $U = (u_1, \ldots, u_s) \in \mathcal{O}_*^s$. Then $R \in V$; and by the Taylor expansion at P,

$$h(R) = h(P) + \pi_*^{\nu+\ell} \sum_{i=1}^{s} \left(\pi_*^{-\ell} \frac{\partial h}{\partial x_i}(P) \right) u_i + \pi_*^{2\nu} j(U) \tag{2.28}$$

where $j(U) \in \mathcal{O}_*[u_1, \ldots, u_s]$. For $\nu \geq \ell$, this equation shows that $h(R) \equiv h(P) \pmod{\pi_*^{\nu+\ell}}$ for all values of $U \in \mathcal{O}_*^s$. Letting U vary modulo π_*^ℓ, we see that

$$\#M_{t,V}(\nu+\ell) = q_*^{s\ell} \cdot \#Y_{t,V}(\nu). \tag{2.29}$$

For what values of U modulo π_* will $\mathbf{red}_{\nu+1}(R) \in Y_{t,V}(\nu+1)$? For $\nu \geq \max\{\nu_0, \ell+1\}$, the answer to this question is clear if we write (2.28) as a congruence modulo $\pi_*^{\nu+1+\ell}$. We find that $\mathbf{red}_{\nu+1}(R) \in Y_{t,V}(\nu+1)$ if and only if

$$t \equiv h(P) + \pi_*^{\nu+\ell} \sum_{i=1}^{s} \left(\pi_*^{-\ell} \frac{\partial h}{\partial x_i}(P) \right) u_i \pmod{\pi_*^{\nu+1+\ell}}.$$

Since $h(P) \equiv t \pmod{\pi_*^{\nu+\ell}}$ and since the coefficient of at least one u_i in the summation on the right is a unit, there are exactly q_*^{s-1} solutions U to this congruence modulo π_*. Thus, each $\bar{P} \in Y_{t,V}(\nu)$ is the image of exactly q_*^{s-1} points in $Y_{t,V}(\nu+1)$. Therefore, for all $\nu \geq \gamma$,

$$\#Y_{t,V}(\nu-\ell) = q_*^{(s-1)(\nu-\gamma)} \cdot \#Y_{t,V}(\gamma-\ell).$$

We obtain (2.27) after multiplying this last equation by $q_*^{s\ell}$ and substituting from (2.29). ■

As a corollary of the proof of this lemma, we get the standard criterion for liftability (cf. Borevich and Shafarevich 1966, Theorem 3 of §5.2):

Proposition 2.9. *Suppose $P \in \mathcal{O}_*^s$ satisfies $h(P) \equiv t \pmod{\pi_*^{2\ell+1}}$, where $\ell = F(P) < +\infty$. Then $\mathbf{red}_{\ell+1}(P)$ is liftable to a point $R \in M_t$.*

Proof. If we take $V = P + \pi_*^{\ell+1}\mathcal{O}_*^s$ in Lemma 2.8, then $2\ell+1$ is the minimum value for γ. As the proof of the lemma shows, every point of $Y_{t,V}(\nu)$ for $\nu \geq \ell+1$ comes by reduction from a point of $Y_{t,V}(\nu+1)$. Put $R_{\ell+1} = P$ and define $R_{\nu+1}$ for $\nu \geq \ell+1$ as any point in V with $\mathbf{red}_{\nu+1}(R_{\nu+1}) \in Y_{t,V}(\nu+1)$ and such that $R_{\nu+1} \equiv R_\nu \pmod{\pi_*^\nu}$. If R is the limit in V of R_ν as $\nu \to \infty$, then $R \in M_{t,V}$ and $\mathbf{red}_{\ell+1}(R) = \mathbf{red}_{\ell+1}(P)$. ■

As we may now prove:

Theorem 2.10. *Suppose that the derivative $\theta_*^u(t;h,V)$ exists as a continuous function on $\mathcal{O}_*$. Suppose further that the singular set $S(h,V)$ of h on V is empty. Put $\delta = \max\{\ell+\nu_0, 2\ell+1\}$, where $\ell = \ell(h,V)$ is defined by* (2.26). *Then*

$$\theta_*^u(t;h,V) = q_*^{(1-s)\delta} \cdot \#M_{t,V}(\delta). \tag{2.30}$$

Proof. By Lemma 2.7, the function (2.25) takes on only a finite number of values on V. Let those values be $\ell_1, \ell_2, \ldots, \ell_r$, and for $1 \le j \le r$, put $V_j = V \cap F^{-1}(\ell_j)$. The sets V_j form a disjoint cover of V by compact sets, which implies that each V_j is also open. Further for $\nu \ge \nu_0$, the union

$$V(\nu) = \bigcup_{j=1}^{r} V_j(\nu)$$

is disjoint, so that

$$\#M_{t,V}(\nu) = \sum_{j=1}^{r} \#M_{t,V_j}(\nu).$$

Now from its definition, $\ell(h,V) = \max\{\ell_j\}$. We may therefore apply Lemma 2.8 to each V_j with $\gamma = \delta$. Summing (2.27) over j with V replaced by V_j and γ by δ, we conclude that

$$\#M_{t,V}(\nu) = q_*^{(s-1)(\nu-\delta)} \cdot \#M_{t,V}(\delta)$$

for all $\nu \ge \delta$. Thus

$$q_*^{(1-s)\nu} \cdot \#M_{t,V}(\nu) = q_*^{(1-s)\delta} \cdot \#M_{t,V}(\delta)$$

for all $\nu \ge \delta$; and this shows that the limit in (2.22) is stationary with value equal to the right hand side of (2.30) for $\nu \ge \delta$. ∎

Exercises 2.5

1. Let $h(X_1,\ldots,X_s) = X_1 + \ldots + X_s$, and suppose $\theta_*(t;h,\mathcal{O}_*^s)$ exists as a continuous function on $\mathcal{O}_*$. Prove the change of variables formula

$$\int_{\mathcal{O}_*^s} f(t_1+\ldots+t_s)\,d_*t_1\cdots d_*t_s = \int_{\mathcal{O}_*} f(t)\,d_*t\,.$$

2. With h as in the previous exercise, suppose (as we prove in §3.3 below) that $\theta_*(t;h,U_*^s)$ exists as a continuous function on $\mathcal{O}_*$. Use Theorem 2.10 to derive the evaluation (3.12).

3. Let p^θ be the largest power of the prime p which divides the positive integer d. Put $\phi = \theta + 2$ if $p = 2$, and put $\phi = \theta + 1$ if $p > 2$. Prove (Hardy and Littlewood 1922b, Lemma 1): if $x = \xi + \alpha p^{\lambda-\theta-1}$ with $\lambda > \phi$, then

$$x^d \equiv \xi^d + \frac{d}{p^\theta}\alpha\xi^{d-1}p^{\lambda-1} \pmod{p^\lambda}.$$

3

Local Gauss sums and local derivatives

In the first two sections of this chapter, we define local Gauss sums as integrals in Haar measure, and we prove some basic theorems evaluating their absolute values. The methods of proof are classical. In the last two sections, we apply these evaluations to compute some local Radon–Nikodym derivatives. The notations introduced in Chapter 2 remain in force.

3.1 A Gauss sum on k_*: definition and properties

Let $d_*^\times u$ be multiplicative Haar measure on $\mathbf{U}_*$ normalized so that the total mass of $\mathbf{U}_*$ is one. By (2.16), the restriction of d_*t to $\mathbf{U}_*$ is also a Haar measure on $\mathbf{U}_*$, and so we have $d_*^\times u = c\, d_*t$ for some constant $c > 0$. Since the d_*t-measure of $\mathbf{U}_*$ is $1 - 1/q_*$,

$$d_*^\times u = \frac{q_*}{q_* - 1}\, d_*t\,. \tag{3.1}$$

Now let Λ be an additive character of k_*^+, and let χ be a multiplicative character of $\mathbf{U}_*$. We set

$$G(\chi, \Lambda) \stackrel{\text{def}}{=} \int_{\mathbf{U}_*} \chi(u)\Lambda(u)\, d_*^\times u\,. \tag{3.2}$$

The integral $G(\chi, \Lambda)$ is called a *normalized Gauss sum on k_**. For convenience, we will usually drop the adjective 'normalized' and refer to $G(\chi, \Lambda)$ as a 'Gauss sum.' We may regard $G(\chi, \Lambda)$ as the value at χ of the Fourier transform of $\Lambda(u)$ as a function on the compact group $\mathbf{U}_*$.

Our calculations in Chapter 6 will require detailed information about how $|G(\chi, \Lambda)|$ varies with the conductors of χ and Λ. The local results we will need are contained in Theorems 3.3 and 3.4 below. The proofs of both theorems are based on the following standard lemma.

Lemma 3.1. *Let G be a compact abelian group with Haar measure dt. Then for all $\chi \in \widehat{G}$,*

$$\int_G \chi(t)\,dt = \begin{cases} 0 & \text{if } \chi \neq \chi_0 \\ meas(G) & \text{if } \chi = \chi_0. \end{cases} \tag{3.3}$$

Proof. If $\chi = \chi_0$, the result is immediate. Assume then that t_0 is an element of G such that $\chi(t_0) \neq 1$. By the invariance of the measure,

$$\int_G \chi(t)\,dt = \int_G \chi(t_0 t)\,dt = \chi(t_0)\int_G \chi(t)\,dt$$

which proves (3.3) when $\chi \neq \chi_0$. ■

Corollary 3.2. *If G is a finite group, then for all $\chi \in \widehat{G}$*

$$\sum_{t \in G} \chi(t) = \begin{cases} 0 & \text{if } \chi \neq \chi_0 \\ \#G & \text{if } \chi = \chi_0. \end{cases}$$

Theorem 3.3. *Suppose that $m(\chi) = 0$, i.e. that χ is the principal character on $\mathbf{U}_*$. Then*

$$G(\chi, \Lambda) = \begin{cases} 1 & \text{if } n(\Lambda) \leq 0 \\ 1/(1 - q_*) & \text{if } n(\Lambda) = 1 \\ 0 & \text{if } n(\Lambda) > 1. \end{cases}$$

Proof. By (3.1), we have

$$\begin{aligned} G(\chi, \Lambda) &= \int_{\mathbf{U}_*} \Lambda(u)\,d_*^\times u = \frac{q_*}{q_* - 1}\int_{\mathbf{U}_*} \Lambda(t)\,d_* t \\ &= \frac{q_*}{q_* - 1}\left(\int_{\mathcal{O}_*} \Lambda(t)\,d_* t - \int_{\pi_*\mathcal{O}_*} \Lambda(t)\,d_* t\right). \end{aligned}$$

If $n(\Lambda) > 1$, then Λ is a non-principal character on both $\mathcal{O}_*$ and $\pi_*\mathcal{O}_*$, and so $G(\chi, \Lambda) = 0$ by (3.3). We leave the remaining two cases to the reader. ■

Theorem 3.4. *Suppose that $m(\chi) \geq 1$. Then*

$$|G(\chi, \Lambda)| = \begin{cases} \frac{q_*^{n/2}}{q_*^{n-1}(q_* - 1)} & \text{if } n(\Lambda) = m(\chi) = n \\ 0 & \text{if } n(\Lambda) \neq m(\chi). \end{cases}$$

Proof. If z is a complex number, we denote its conjugate by $\bar{z}$. Using the invariance of the Haar measure and Fubini's theorem, we compute

$$\begin{aligned}|G(\chi,\Lambda)|^2 &= G(\chi,\Lambda)\cdot\overline{G(\chi,\Lambda)} \\ &= \int_{\mathbf{U}_*}\int_{\mathbf{U}_*}\chi(u)\overline{\chi(v)}\Lambda(u)\Lambda(-v)\,d_*^\times u\,d_*^\times v \\ &= \int_{\mathbf{U}_*}\int_{\mathbf{U}_*}\chi(uv)\overline{\chi(v)}\Lambda(uv-v)\,d_*^\times u\,d_*^\times v \\ &= \int_{\mathbf{U}_*}\chi(u)\left(\int_{\mathbf{U}_*}\Lambda^{u-1}(v)\,d_*^\times v\right)d_*^\times u\,.\end{aligned}$$

Applying Theorem 3.3 to the inner integral and adopting the shorthand notation '$n(\Lambda^{u-1})\le 0$' for '$\{u\in\mathbf{U}_*: n(\Lambda^{u-1})\le 0\}$', etc., we compute

$$\begin{aligned}|G(\chi,\Lambda)|^2 &= \int_{n(\Lambda^{u-1})\le 0}\chi(u)\,d_*^\times u-\frac{1}{q_*-1}\int_{n(\Lambda^{u-1})=1}\chi(u)\,d_*^\times u \\ &= \frac{q_*}{q_*-1}\int_{n(\Lambda^{u-1})\le 0}\chi(u)\,d_*^\times u-\frac{1}{q_*-1}\int_{n(\Lambda^{u-1})\le 1}\chi(u)\,d_*^\times u\,.\end{aligned}$$

By (2.12), $n(\Lambda^{u-1})=n(\Lambda)-\mathrm{v}_*(u-1)$, so $n(\Lambda^{u-1})\le 1$ if and only if $\mathrm{v}_*(u-1)\ge n(\Lambda)-1$; and $n(\Lambda^{u-1})\le 0$ if and only if $\mathrm{v}_*(u-1)\ge n(\Lambda)$. Setting $n=n(\Lambda)$, we observe that $\{u: n(\Lambda^{u-1})\le 1\}=\mathbf{U}_*^{(n-1)}$ and $\{u: n(\Lambda^{u-1})\le 0\}=\mathbf{U}_*^{(n)}$. Thus

$$|G(\chi,\Lambda)|^2=\frac{1}{q_*-1}\left(q_*\int_{\mathbf{U}_*^{(n)}}\chi(u)\,d_*^\times u-\int_{\mathbf{U}_*^{(n-1)}}\chi(u)\,d_*^\times u\right). \tag{3.4}$$

It remains then to evaluate the two integrals on the right above. We consider three cases:

Case 1. Suppose $n=n(\Lambda)<m(\chi)$. Then χ is non-trivial on both $\mathbf{U}_*^{(n)}$ and $\mathbf{U}_*^{(n-1)}$, so both integrals are zero by (3.3).

Case 2. Suppose that $n(\Lambda)=m(\chi)=n$. Then χ is trivial on $\mathbf{U}_*^{(n)}$ but non-trivial on $\mathbf{U}_*^{(n-1)}$. For $\ell\ge 1$, the $d_*^\times$-measure of $\mathbf{U}_*^{(\ell)}$ is $1/(q_*^{\ell-1}(q_*-1))$. Together with (3.4) and (3.3), this evaluation shows that

$$|G(\chi,\Lambda)|^2=\frac{q_*}{q_*-1}\cdot\frac{1}{q_*^{n-1}(q_*-1)}-0=\frac{q_*^n}{q_*^{2n-2}(q_*-1)^2}\,.$$

Taking the square root of both sides, we get the desired result.

Case 3. Suppose $n=n(\Lambda)>m(\chi)$. Then χ is trivial on both $\mathbf{U}_*^{(n)}$ and $\mathbf{U}_*^{(n-1)}$. But since $meas(\mathbf{U}_*^{(n-1)})=q_*\cdot meas(\mathbf{U}_*^{(n)})$, (3.4) implies that $G(\chi,\Lambda)=0$. ∎

Theorem 3.4 says that if χ is non-trivial on $\mathbf{U}_*$ and if χ and Λ 'match up' (in the sense that the conductors are equal), then the modulus of the corresponding Gauss sum is $q_*^{n/2}$ times the $d_*^\times$-measure of $\mathbf{U}_*^{(n)}$. A similar result will emerge in Theorem 3.7 below.

The classical evaluations of the absolute values of Gauss sums on $\mathbb{Z}/p^\ell\mathbb{Z}$, where p is a prime number and $\ell \geq 0$, are special cases of the results in this section.

Exercises 3.1

1. Let **red** : $\mathcal{O}_* \longrightarrow \mathbf{K}(*)$ be the canonical map 'reduction modulo π_*'. Prove that the $d_*^\times$-measure of $\mathbf{U}_*^{(\ell)}$ is $1/(q_*^{\ell-1}(q_*-1))$ for all $\ell \geq 1$ as follows:

   (a) Show that $\mathbf{U}_*/\mathbf{U}_*^{(1)} \cong \mathbf{K}(*)^\times$ by looking at the restriction of **red** to $\mathbf{U}_*$.

   (b) Show that for $\ell \geq 1$, $\mathbf{U}_*^{(\ell)}/\mathbf{U}_*^{(\ell+1)} \cong \mathbf{K}(*)^+$ by looking at the function which maps $1+\pi_*^\ell t \in \mathbf{U}_*^{(\ell)}$ to $\mathbf{red}(t)$.

   (c) Conclude that $\#(\mathbf{U}_*/\mathbf{U}_*^{(\ell)}) = q_*^{\ell-1}(q_*-1)$, and hence the result.

2. Let r be an integer greater than or equal to both $n(\Lambda)$ and $m(\chi)$, and let b run through any system of representatives for the cosets of $\mathbf{U}_*^{(r)}$ in $\mathbf{U}_*$. Show that

$$G(\chi,\Lambda) = \frac{1}{q_*^{r-1}(q_*-1)} \sum_b \chi(b)\Lambda(b) .$$

3.2 A modified Gauss sum

In this section, we study another Gauss sum whose properties are similar to those of $G(\chi,\Lambda)$. We start by renormalizing the multiplicative Haar measure $d_*^\times u$ as follows: let $d_*^{(1)}u \stackrel{\text{def}}{=} (q_*-1)\cdot d_*^\times u$. Then since $\#(\mathbf{U}_*/\mathbf{U}_*^{(1)}) = q_*-1$, $d_*^{(1)}u$ is the Haar measure for which $\mathbf{U}_*^{(1)}$ gets measure one. Now define

$$G^{(1)}(\chi,\Lambda) \stackrel{\text{def}}{=} \int_{\mathbf{U}_*^{(1)}} \chi(u)\Lambda(u)\, d_*^{(1)}u \tag{3.5}$$

where $\chi \in \widehat{\mathbf{U}}_*^{(1)}$ and $\Lambda \in \hat{k}_*^+$. We call $G^{(1)}(\chi,\Lambda)$ a *modified Gauss sum*. The analogue of Theorem 3.3 becomes:

Theorem 3.5. *Suppose that χ is trivial on $\mathbf{U}_*^{(1)}$. Then*

$$G^{(1)}(\chi, \Lambda) = \begin{cases} \Lambda(1) & \text{if } n(\Lambda) \leq 1 \\ 0 & \text{if } n(\Lambda) > 1. \end{cases}$$

Proof. We have

$$\begin{aligned} G^{(1)}(\chi, \Lambda) &= \int_{\mathbf{U}_*^{(1)}} \Lambda(u)\, d_*^{(1)}u = q_* \int_{\pi_* \mathcal{O}_*} \Lambda(1+t)\, d_*t \\ &= q_* \Lambda(1) \int_{\pi_* \mathcal{O}_*} \Lambda(t)\, d_*t = \begin{cases} \Lambda(1) & \text{if } n(\Lambda) \leq 1 \\ 0 & \text{if } n(\Lambda) > 1 \end{cases} \end{aligned}$$

by (3.3). ■

The evaluation of $|G^{(1)}(\chi, \Lambda)|$ when χ is non-trivial is more complicated than was the case for $G(\chi, \Lambda)$. By Exercise 3.1(1), $\mathbf{U}_*/\mathbf{U}_*^{(1)} \cong \mathbf{K}(*)^\times$, which is a cyclic group of order $q_* - 1$. Therefore $\mathbf{U}_* = \mu_{q_*-1} \cdot \mathbf{U}_*^{(1)}$, where μ_{q_*-1} is the multiplicative group of $(q_* - 1)$-st roots of unity in $\mathbf{U}_*$. Hence, we can extend any $\chi \in \widehat{\mathbf{U}}_*^{(1)}$ to a character of $\mathbf{U}_*$ in exactly $q_* - 1$ ways. Fix one of these extensions, and call it 'χ' also. Then all extensions are of the form $\psi\chi$ with $\psi \in \hat{\mu}_{q_*-1}$. Writing $v \in \mathbf{U}_*$ as $v = zu$ with $z \in \mu_{q_*-1}$ and $u \in \mathbf{U}_*^{(1)}$, we have

$$\begin{aligned} G(\psi\chi, \Lambda) &= \int_{\mathbf{U}_*} (\psi\chi)(v)\Lambda(v)\, d_*^\times v \\ &= \int_{\mu_{q_*-1}} \int_{\mathbf{U}_*^{(1)}} \psi(z)\chi(u)\Lambda(zu)\, d_*^\times u\, dz \\ &= \frac{1}{q_* - 1} \int_{\mu_{q_*-1}} \psi(z) \left(\int_{\mathbf{U}_*^{(1)}} \chi(u)\Lambda(zu)\, d_*^{(1)}u \right) dz , \end{aligned}$$

where dz is counting measure on the finite group μ_{q_*-1}. Now with ψ ranging over $\hat{\mu}_{q_*-1}$, we have

$$\sum_\psi \psi(z) = \begin{cases} 0 & \text{if } z \neq 1 \\ q_* - 1 & \text{if } z = 1 \end{cases}$$

by Corollary 3.2; and so

$$G^{(1)}(\chi, \Lambda) = \sum_\psi G(\psi\chi, \Lambda) . \tag{3.6}$$

We are now in a position to calculate $|G^{(1)}(\chi, \Lambda)|$.

Theorem 3.6. *Suppose that χ is not trivial on $\mathbf{U}_*^{(1)}$. If $n(\Lambda) \neq m(\chi)$, then $G^{(1)}(\chi, \Lambda) = 0$.*

Proof. The extensions of χ to $\mathbf{U}_*$ are all non-principal because they each restrict to χ on $\mathbf{U}_*^{(1)}$. The result follows now by (3.6) and Theorem 3.4. ∎

We cannot say, as we can in the case of $G(\chi, \Lambda)$, that $|G^{(1)}(\chi, \Lambda)|$ has a certain value provided that the conductors of χ and Λ are equal. Let us assume that $n(\Lambda) = m(\chi) = n \geq 2$. By restriction, χ induces a character $\chi^\natural$ on the quotient $\mathbf{U}_*^{(n-1)}/\mathbf{U}_*^{(n)}$ and Λ induces a character $\Lambda^\natural$ on the quotient $\pi_*^{n-1}\mathcal{O}_*/\pi_*^n\mathcal{O}_*$. Moreover, the mapping $u \mapsto u - 1$ lifts to a group isomorphism $\eta : \mathbf{U}_*^{(n-1)}/\mathbf{U}^{(n)} \longrightarrow \pi_*^{n-1}\mathcal{O}_*/\pi_*^n\mathcal{O}_*$. If

$$\Lambda^\natural \circ \eta = \overline{\chi^\natural}\,, \tag{3.7}$$

then we say that χ and Λ are *n-cancellative.* We can now prove:

Theorem 3.7. *Suppose $n(\Lambda) = m(\chi) \geq 2$. Then*

$$|G^{(1)}(\chi, \Lambda)| = \begin{cases} \frac{q_*^{n/2}}{q_*^{n-1}} & \textit{if } \chi \textit{ and } \Lambda \textit{ are } n\textit{-cancellative} \\ 0 & \textit{otherwise}\,. \end{cases}$$

Proof. We have

$$\begin{aligned} |G^{(1)}(\chi, \Lambda)|^2 &= \int_{\mathbf{U}_*^{(1)}} \int_{\mathbf{U}_*^{(1)}} \chi(u)\overline{\chi(v)}\Lambda(u)\Lambda(-v)\, d_*^{(1)}u\, d_*^{(1)}v \\ &= \int_{\mathbf{U}_*^{(1)}} \int_{\mathbf{U}_*^{(1)}} \chi(uv)\overline{\chi(v)}\Lambda(uv - v)\, d_*^{(1)}u\, d_*^{(1)}v \\ &= \int_{\mathbf{U}_*^{(1)}} \chi(u) \left(\int_{\mathbf{U}_*^{(1)}} \Lambda^{u-1}(v)\, d_*^{(1)}v \right) d_*^{(1)}u\,. \end{aligned}$$

By Theorem 3.5, the inner integral above equals either 0 or $\Lambda^{(u-1)}(1) = \Lambda(u-1)$ depending on the value of the conductor of $\Lambda^{(u-1)}$. For $\mathrm{u} \in \mathbf{U}_*^{(1)}$, $n(\Lambda^{(\mathrm{u}-1)}) \leq 1$ is equivalent to $n - \mathrm{v}_*(\mathrm{u}-1) \leq 1$ by (2.12) which in turn holds if and only if $\mathrm{u} \in \mathbf{U}_*^{(n-1)}$. Therefore

$$|G^{(1)}(\chi, \Lambda)|^2 = \int_{\mathbf{U}_*^{n-1}} \chi(\mathrm{u})\Lambda(\mathrm{u}-1)\, d_*^{(1)}\mathrm{u}\,.$$

Let z run through the cosets of $\mathbf{U}_*^{(n)}$ in $\mathbf{U}_*^{(n-1)}$. Since these cosets partition $\mathbf{U}_*^{(n-1)}$, the integral on the right above equals

$$\frac{1}{q_*^{n-1}} \cdot \sum_z \chi^\natural(z) \cdot (\Lambda^\natural \circ \eta)(z) = \frac{1}{q_*^{n-1}} \begin{cases} q_* & \text{for } \chi, \Lambda\text{-}n\text{-cancellative} \\ 0 & \text{otherwise} \end{cases}$$

because the $d_*^{(1)}$-measure of $\mathbf{U}_*^{(n)}$ equals $1/q_*^{n-1}$. Taking square roots, we obtain our result. ∎

We observe in analogy with Theorem 3.4, that when χ and Λ 'match up', $|G^{(1)}(\chi, \Lambda)|$ is $q_*^{n/2}$ times the $d_*^{(1)}$-measure of $\mathbf{U}_*^{(n)}$.

3.3 Computation of some special local derivatives

In this section, we will use our results from §3.1 to compute some examples of Radon–Nikodym derivatives $\theta_*(t; h, V)$ on $\mathcal{O}_*$ (cf. §2.4). These computations provide us with the local factors in the asymptotic solutions of the 3-primes and Waring's problem in $\mathbf{A}_q$. Since we do not wish to appeal to the general theory outlined in the last paragraphs of §2.4, we will use a Fourier series method to prove that derivatives exist for our examples. The Fourier series also reveals important additional information about the derivatives. For example, we will obtain upper and lower bounds for their absolute values. This method has a venerable history since it was used by Hardy and Littlewood in their original papers on Waring's problem and the classical 3-primes problem. The Fourier series we compute are local versions of the Hardy–Littlewood singular series.

The efficient implementation of the Fourier series method requires a basic result from harmonic analysis. Let $\mathcal{C}(G)$ be the space of continuous $\mathbb{C}$-valued functions on the compact abelian group G. For $f \in \mathcal{C}(G)$,

$$\|f\| = \sup_{x \in G} |f(x)| \tag{3.8}$$

is called the *sup-norm* of f; and one knows that $\mathcal{C}(G)$ is a Banach algebra in this norm. The characters of G are functions in $\mathcal{C}(G)$ of norm one. Let W be the $\mathbb{C}$-vector subspace of $\mathcal{C}(G)$ consisting of the finite linear combinations of characters in $\widehat{G}$. Then we have

Theorem 3.8. *The subspace W is a dense subset of $\mathcal{C}(G)$ in the topology induced by the norm* (3.8).

Proof. This theorem is a consequence of the Stone–Weierstrass theorem (see Loomis 1953, §38). ∎

The topology induced by (3.8) on $\mathcal{C}(G)$ is called the *uniform topology*. Theorem 3.8 shows that a continuous function on G can be uniformly approximated to any degree of accuracy by a finite linear combination of characters. In order to apply this theorem to the computation of derivatives on $\mathcal{O}_*$, we need a character $\Lambda \in \hat{k}_*^+$ with $n(\Lambda) = 0$. We fix such a character for the remainder of this section.

As a first example, we will compute the derivative of the linear polynomial

$$h(X_1, X_2, \ldots, X_s) = X_1 + X_2 + \cdots + X_s \tag{3.9}$$

as a function on $V = \mathbf{U}_*^s$. If the derivative $\theta_*(t; h, V)$ exists, as an L^1-function on $\mathcal{O}_*$, then we may compute its Fourier coefficient (2.18) at $x = a/\pi_*^\nu$ as follows:

$$\begin{aligned} c_h(a/\pi_*^\nu) &= \int_{\mathcal{O}_*} \theta_*(t; h, V)\Lambda(xt)\, d_*t \\ &= \int_{\mathbf{U}_*^s} \Lambda(x(t_1 + t_2 + \cdots + t_s))\, \frac{d_*t_1\, d_*t_2 \cdots d_*t_s}{meas(\mathbf{U}_*^s)} \\ &= [G(\chi_0, \Lambda^x)]^s \end{aligned}$$

from the definitions (3.1) and (3.2). Now $n(\Lambda^x) = n(\Lambda) - \mathrm{v}_*(x) = -\mathrm{v}_*(x) = \nu$. Applying Theorem 3.3, we see that

$$c_h(a/\pi_*^\nu) = \begin{cases} 1 & \text{if } \nu = 0 \\ (1 - q_*)^{-s} & \text{if } \nu = 1 \\ 0 & \text{if } \nu > 1. \end{cases} \tag{3.10}$$

Thus if $\theta_*(t; h, V)$ exists as an L^1-function on $\mathcal{O}_*$, then its Fourier series expansion (2.19) is

$$\theta_*(t; h, V) = 1 + \frac{1}{(1 - q_*)^s} \sum_{a/\pi_*} \Lambda(-at/\pi_*) \tag{3.11}$$

where a/π_* runs through any set of representatives for the non-zero cosets of the quotient group $\pi_*^{-1}\mathcal{O}_*/\mathcal{O}_*$. For fixed t, $\Lambda(-at/\pi_*)$ is a character of this quotient; and therefore the summation on the right in (3.11) equals $q_* - 1$ if this character is trivial and equals -1 otherwise. Thus, we find

$$\theta_*(t; h, V) = \begin{cases} 1 - (1 - q_*)^{-s} & \text{if } \mathrm{v}_*(t) = 0 \\ 1 - (1 - q_*)^{1-s} & \text{if } \mathrm{v}_*(t) > 0 \end{cases} \tag{3.12}$$

provided only that $\theta_*(t; h, V)$ exists!

Let us now define a function $g(t)$ for $t \in \mathcal{O}_*$ by the right hand side of (3.12). Then $g(t)$ is continuous and hence belongs to $L^1(\mathcal{O}_*)$. Further, by its definition, $g(t)$ satisfies

$$\int_{\mathcal{O}_*} f(t)g(t)\, d_*t = \int_{\mathbf{U}_*^s} f(h(t_1, \ldots, t_s))\, \frac{d_*t_1 \cdots d_*t_s}{meas(\mathbf{U}_*^s)} \tag{3.13}$$

whenever $f \in \mathcal{C}(\mathcal{O}_*)$ is a character. Since both sides of (3.13) are continuous linear functions of f in the uniform topology of $\mathcal{C}(\mathcal{O}_*)$, the identity (3.13) must hold for all $f \in \mathcal{C}(\mathcal{O}_*)$ by Theorem 3.8. We have now proved:

Theorem 3.9. *The derivative $\theta_*(t;h,V)$ of the linear polynomial* (3.9) *as a function on $V = \mathbf{U}_*^s$ exists and is continuous on $\mathcal{O}_*$. Further, $\theta_*(t;h,V)$ is given explicitly by* (3.12).

The derivative $\theta_*(t;h,V)$ for $s = 3$ is the local singular series for the 3-primes problem.

We now turn to a second example. Let $d_1, d_2, \dots, d_s$ be strictly positive integers no one of which is divisible by $\mathrm{char}(k_*)$. For given units $\gamma_1, \gamma_2, \dots, \gamma_s \in \mathbf{U}_*$, let

$$h(X_1, X_2, \dots, X_s) = \gamma_1 X_1^{d_1} + \gamma_2 X_2^{d_2} + \cdots + \gamma_s X_s^{d_s}$$

be the corresponding diagonal polynomial. In this example, we take $V = \mathcal{O}_*^s$. We will show that the Radon–Nikodym derivative $\theta_*(t;h,\mathcal{O}_*^s)$ exists as a continuous function on $\mathcal{O}_*$ provided that

$$\frac{1}{d_1} + \frac{1}{d_2} + \cdots + \frac{1}{d_s} > 1\,.$$

Lemma 3.10. *Let $P_d(X) = X^d$ for d a positive integer which is not divisible by* $\mathrm{char}(k_*)$. *Then $P_d(\mathbf{U}_*)$ is an open subgroup of $\mathbf{U}_*$. Further, for all $f \in L^1(d_*u)$*

$$\int_{\mathbf{U}_*} f(u^d)\, d_*u = c_*(d) \int_{P_d(\mathbf{U}_*)} f(w)\, d_*w$$

where $c_*(d) = \#\mathbf{U}_*/P_d(\mathbf{U}_*)$ is the index in $\mathbf{U}_*$ of its subgroup of d-th powers.

Proof. Using Proposition 2.9 with $s = 1$, we can prove that $P_d(\mathbf{U}_*)$ contains one of the groups (2.8) and is therefore an open subgroup of $\mathbf{U}_*$. The measure du_* projects under the map P_d to a measure ρ on $P_d(\mathbf{U}_*)$. By definition,

$$\int_{P_d(\mathbf{U}_*)} f(w)\, d\rho(w) = \int_{\mathbf{U}_*} f(u^d)\, d_*u \tag{3.14}$$

for every continuous function f on $\mathbf{U}_*$. It is clear from (3.14) that ρ is a Haar measure on the group $P_d(\mathbf{U}_*)$. Therefore $d\rho(w) = c_*(d)\, d_*w$ for a positive constant $c_*(d)$. To evaluate this constant, we take $f(w) = 1$ in (3.14) and conclude that

$$c_*(d) = meas(\mathbf{U}_*)/meas(P_d(\mathbf{U}_*)) = \#\mathbf{U}_*/P_d(\mathbf{U}_*)\,. \tag{3.15}$$

■

Lemma 3.11. *For $z \in k_*^\times$, let*

$$\tau_d(z) = \int_{\mathcal{O}_*} \Lambda(zt^d)\, d_*t\,.$$

If $\mathrm{v}_*(z) = -\nu$ with $\nu \geq 0$, then

$$|\tau_d(z)| \leq c_*(d) q_*^{-\nu/d} \tag{3.16}$$

where $c_*(d)$ is the integer defined by (3.15).

Proof. We have

$$\tau_d(z) = \int_{\mathbf{U}_*} \Lambda(zt^d)\, d_*t + \int_{\pi_*\mathcal{O}_*} \Lambda(zt^d)\, d_*t\,.$$

The second integral equals $q_*^{-1}\cdot\tau_d(\pi_*^d z)$ after the change of variable $t \leftarrow \pi_* t$. Therefore

$$\tau_d(z) - q_*^{-1}\tau_d(\pi_*^d z) = \int_{\mathbf{U}_*} \Lambda(zt^d)\, d_*t\,. \tag{3.17}$$

By Lemma 3.10, the integral on the right in (3.17) is

$$c_*(d) \cdot \int_{P_d(\mathbf{U}_*)} \Lambda(zw)\, d_*w\,. \tag{3.18}$$

Let $H \subset \widehat{\mathbf{U}}_*$ be the subgroup consisting of the characters χ such that $\chi^d = 1$. By (3.15), there are precisely $c_*(d)$ such characters. Therefore

$$\frac{1}{c_*(d)} \sum_{\chi \in H} \chi(w)$$

is the characteristic function of $P_d(\mathbf{U}_*)$ in $\mathbf{U}_*$. This follows from (3.3) if we take $G = H$ and integrate the character $\widehat{w} \in \widehat{H}$ defined for fixed $w \in \mathbf{U}_*$ by $\widehat{w}(\chi) = \chi(w)$. Now we can write (3.18) as

$$\sum_{\chi \in H} \int_{\mathbf{U}_*} \Lambda(zw)\chi(w)\, d_*w$$

which implies via (3.17) that

$$\tau_d(z) - q_*^{-1}\tau_d(\pi_*^d z) = \sum_{\chi \in H} \int_{\mathbf{U}_*} \Lambda(zw)\chi(w)\, d_*w\,.$$

In order to compute $\tau_d(z)$ itself, we observe that

$$\tau_d(z) \;=\; \sum_{\lambda=0}^{\infty} q_*^{-\lambda} \left[\tau_d(\pi_*^{d\lambda} z) - q_*^{-1} \cdot \tau_d(\pi_*^{d\lambda+d} z)\right]$$

$$= \sum_{\lambda=0}^{\infty} q_*^{-\lambda} \left[\sum_{\chi \in H} \int_{\mathbf{U}_*} \Lambda(\pi_*^{d\lambda} zw)\chi(w)\, d_*w \right]$$

$$= \sum_{\chi \in H} \sum_{\lambda=0}^{\infty} q_*^{-\lambda} \int_{\mathbf{U}_*} \Lambda(\pi_*^{d\lambda} zw)\chi(w)\, d_*w\ . \tag{3.19}$$

Let Σ_0 be the contribution of the principal character to (3.19), and let Σ_1 be the remainder of the summation over χ. Each integral in the second sum Σ_1 is a Gauss sum involving an additive character of conductor $-\mathrm{v}_*(z) - d\lambda = \nu - d\lambda$ (by (2.12) since $n(\Lambda) = 0$) and a non-principal multiplicative character χ. If $m = m(\chi)$, all the integrals involving χ are *zero* except for the one with $\nu - d\lambda = m$ by Theorem 3.4. By this same theorem together with (3.1), the non-zero integral involving χ has absolute value $q_*^{-m/2}$. Therefore,

$$|\Sigma_1| \le \sum_{\chi \ne \chi_0} q_*^{\frac{m-\nu}{d}} q_*^{-m/2} \le (c_*(d) - 1) \cdot q_*^{-\nu/d} \tag{3.20}$$

as Σ_1 is empty unless $d \ge 2$.

Turning now to the sum Σ_0, we first assume $\nu \equiv 1 \pmod d$. The integrals appearing in Σ_0 can be evaluated by Theorem 3.3 together with (3.1). We find that

$$\begin{aligned} \Sigma_0 &= -q_*^{\frac{1-\nu}{d}} \frac{1}{q_*} + \sum_{\lambda \ge \nu/d} q_*^{-\lambda} \left[1 - \frac{1}{q_*} \right] \\ &= q_*^{-\{\nu/d\}} - q_*^{-1+(1-\nu)/d} \end{aligned}$$

where $\{\nu/d\}$ is the least integer greater than or equal to ν/d. Since $\nu/d = (\nu-1)/d + 1/d$, we have $\{\nu/d\} = (\nu-1)/d + 1$ so that $\Sigma_0 = 0$ in this case. Now if $\nu \not\equiv 1 \pmod d$, we make the same computation except that there is no negative contribution from a term with $\nu - d\lambda = 1$. We conclude that

$$\Sigma_0 = q_*^{-\{\nu/d\}} \le q_*^{-\nu/d}\ .$$

Of course, this last inequality is also valid when $\nu \equiv 1 \pmod d$. Substituting it and the bound (3.20) into (3.19), we obtain (3.16). ∎

If the derivative $\theta_*(t; h, \mathcal{O}_*^s)$ exists as an L^1-function on $\mathcal{O}_*$, then we may compute its Fourier coefficient (2.18) at $x = a/\pi_*^\nu$ as follows:

$$c_h(a/\pi_*^\nu) = \int_{\mathcal{O}_*} \theta_*(t; h, \mathcal{O}_*^s)\Lambda(xt)\, d_*t$$

$$= \int_{\mathcal{O}_*^s} \Lambda(xh(t_1,\ldots,t_s))\, d_*t_1 d_*t_2 \cdots d_*t_s \tag{3.21}$$

$$= \prod_{i=1}^{s} \tau_{d_i}(x\gamma_i)\,.$$

Therefore if $\theta_*(t; h, \mathcal{O}_*^s)$ exists, then it has the formal Fourier expansion

$$\theta_*(t; h, \mathcal{O}_*^s) = \sum_{\nu=0}^{\infty} \sum_{a/\pi_*^\nu} c_h(a/\pi_*^\nu)\Lambda(-at/\pi_*^\nu) \tag{3.22}$$

where for each ν, a/π_*^ν runs through a set of representatives for the primitive cosets in $\pi_*^{-\nu}\mathcal{O}_*/\mathcal{O}_*$. Let d be the rational number defined by

$$\frac{s}{d} = \frac{1}{d_1} + \frac{1}{d_2} + \cdots + \frac{1}{d_s}\,.$$

Appealing to Lemma 3.11, we find the bound

$$|c_h(a/\pi_*^\nu)| \le \prod_{i=1}^{s} c_*(d_i) \cdot q_*^{-\nu/d_i} = C_*(h) \cdot q_*^{-\nu s/d} \tag{3.23}$$

where

$$C_*(h) = \prod_{i=1}^{s} c_*(d_i)\,.$$

We may now prove the following.

Theorem 3.12. *If $s > d$, then the derivative $\theta_*(t; h, \mathcal{O}_*^s)$ exists as a continuous function on $\mathcal{O}_*$. Further, the Fourier expansion* (3.22) *converges absolutely and uniformly to $\theta_*(t; h, \mathcal{O}_*^s)$ for every $t \in \mathcal{O}_*$.*

Proof. For each $\nu \ge 0$,

$$\sum_{a/\pi_*^\nu} |c_h(a/\pi_*^\nu)| \le C_*(h) \cdot q_*^{\nu(1-s/d)} \tag{3.24}$$

by (3.23) since the number of elements in the summation is less than q_*^ν. Since $s > d$, this shows that the series in (3.22) is absolutely and uniformly convergent for $t \in \mathcal{O}_*$. Let $g(t)$ be the sum of that series. Then $g(t)$ is a continuous function on $\mathcal{O}_*$. Using Theorem 3.8 as we did in proving Theorem 3.9 above, we see that $g(t)$ effects the change of variables for the projection of the product measure on $\mathcal{O}_*^s$ to $\mathcal{O}_*$ via h. Therefore $\theta_*(t; h, \mathcal{O}_*^s) = g(t)$ as required. ■

Lemma 3.11 also implies bounds for the magnitude of $\theta_*(t; h, \mathcal{O}_*^s)$. Suppose $d > 0$ is not divisible by the residual characteristic char($\mathbf{K}(*)$) of k_*. Then Proposition 2.9 with $s = 1$ implies that $P_d(\mathbf{U}_*)$ contains $\mathbf{U}_*^{(1)}$. Since $\mathbf{U}_*/\mathbf{U}_*^{(1)}$ is isomorphic to the multiplicative group of $\mathbf{K}(*)$, the quotient $\mathbf{U}_*/P_d(\mathbf{U}_*)$ has the cardinality of the kernel of the d-th power map on $\mathbf{K}(*)^\times$. Therefore, $c_*(d) = \mu_d(\mathbf{K}(*)) \leq d$.

Theorem 3.13. *Assume that* char($\mathbf{K}(*)$) *does not divide any of the exponents* $d_1, d_2, \ldots, d_s$. *Assume further that* $s > 2d$. *Then*

$$|\theta_*(t; h, \mathcal{O}_*^s) - 1| \leq 2d_1 d_2 \ldots d_s \cdot q_*^{(1-s/d)}$$

for all $t \in \mathcal{O}_*$.

Proof. From (3.24) and the series expansion (3.22), we compute

$$\begin{aligned} |\theta_*(t; h, \mathcal{O}_*^s) - 1| &\leq \sum_{\nu=1}^{\infty} C_*(h) \cdot q_*^{\nu(1-s/d)} \\ &= C_*(h) \cdot \frac{q_*^{(1-s/d)}}{1 - q_*^{(1-s/d)}} . \end{aligned}$$

Since $s > 2d$, we have $1 - s/d < -1$ so that

$$1 - q_*^{(1-s/d)} > 1 - q_*^{-1} \geq \frac{1}{2} .$$

Also, the hypotheses on the exponents imply that $c_*(d_i) \leq d_i$ for all $1 \leq i \leq s$. Therefore $C_*(h) \leq d_1 d_2 \ldots d_s$. ■

Exercises 3.3

1. Use Exercise 2.4(4) to show, without any hypothesis on the residual characteristic of k_*, that

$$C_*(h) \leq \frac{d_1 d_2 \ldots d_s}{|d_1 d_2 \ldots d_s|_*} .$$

3.4 The local derivative for Waring's problem

Let d be a strictly positive integer which is prime to the residual characteristic $p = \mathrm{char}(\mathbf{K}(*))$, and assume that $s > d$. Let $\gamma_1, \gamma_2, \ldots, \gamma_s$ be fixed elements of $\mathbf{U}_*$. Our aim in this section is to compute the derivative $\theta_*(t; h, \mathcal{O}_*^s)$ of the special polynomial

$$h(X_1, X_2, \ldots, X_s) = \gamma_1 X_1^d + \gamma_2 X_2^d + \cdots + \gamma_s X_s^d \tag{3.25}$$

in a fairly explicit form. Since $s > d$, we know from Theorem 3.12 that $\theta_*(t; h, \mathcal{O}^s)$ exists as a continuous function on $\mathcal{O}_*$.

First, we decompose $\mathcal{O}_*^s$ as the union

$$\mathcal{O}_*^s = V_1 \cup V_2 \tag{3.26}$$

where $V_1 = (\pi_* \mathcal{O}_*)^s$ and $V_2 = \mathcal{O}_*^s - V_1$. Since (3.26) expresses $\mathcal{O}_*^s$ as the disjoint union of two compact open subsets, we have

$$\theta_*(t; h, \mathcal{O}^s) = \theta_*^u(t; h, V_1) + \theta_*^u(t; h, V_2) \tag{3.27}$$

in the sense that if either one of the unnormalized derivatives on the right exists, then the other one also exists and the equality holds. This fact follows from the definitions (see §2.4) in a straightforward manner. We will show by direct means that $\theta_*^u(t; h, V_1)$ exists as a continuous function on $\mathcal{O}_*$. The existence and continuity of $\theta_*^u(t; h, V_2)$ will then follow from (3.27).

Consider $\theta_*(t; h, \mathcal{O}^s)$ to be 'extended by zero' to all of k_*. This means that the domain of $\theta_*(t; h, \mathcal{O}^s)$ is enlarged to k_* by defining $\theta_*(t; h, \mathcal{O}^s) = 0$ for $t \in k_* - \mathcal{O}_*$. Since $\mathcal{O}_*$ is both closed and open in k_*, this extended function remains continuous. Now for any $f \in C(\mathcal{O}_*)$, we have

$$\int_{\mathcal{O}_*} f(\pi_*^d t) \theta_*(t; h, \mathcal{O}^s)\, d_* t = q_*^d \int_{\mathcal{O}_*} f(t) \theta_*(\pi_*^{-d} t; h, \mathcal{O}^s)\, d_* t$$

by (2.16). On the other hand,

$$\begin{aligned} \int_{\mathcal{O}_*} f(\pi_*^d t) \theta_*(t; h, \mathcal{O}^s)\, d_* t &= \int_{\mathcal{O}_*^s} f(\pi_*^d h(t_1, \ldots, t_s))\, d_* t_1 \cdots d_* t_s \\ &= q_*^s \int_{V_1} f(h(t_1, \ldots, t_s))\, d_* t_1 \cdots d_* t_s \end{aligned}$$

by (2.16) together with Fubini's theorem. We conclude from these last two equations that

$$\theta_*^u(t; h, V_1) = q_*^{d-s} \cdot \theta_*(\pi_*^{-d} t; h, \mathcal{O}^s)\,. \tag{3.28}$$

Thus, when $s > d$, $\theta_*^u(t; h, V_1)$ and hence $\theta_*^u(t; h, V_2)$ exist as continuous functions on $\mathcal{O}_*$.

For $t \in \mathcal{O}_*$, let H_t be the set of $\mathbf{K}(*)$-valued points on the hypersurface defined by $h(X_1, \ldots, X_s) = t$ over $\mathcal{O}_*$. Let $N_h(t)$ be the number of non-zero points in H_t. Then $t \notin \pi_*\mathcal{O}_* \Rightarrow N_h(t) = \#H_t$ whereas $t \in \pi_*\mathcal{O}_* \Rightarrow N_h(t) = (\#H_t) - 1$. In the notation introduced in §2.5,

$$N_h(t) = \#M_{t,V_2}(1)$$

because V_2 is the set of points of $\mathcal{O}_*^s$ which have non-zero reduction modulo π_*. Since $p \not| d$ and because every point of V_2 has at least one coordinate in $\mathbf{U}_*$, the function $F(P)$ defined by (2.25) takes the value zero at every point $P \in V_2$. Therefore, Theorem 2.10 with $V = V_2$ and h defined by (3.25) is valid with $\delta = 1$. We conclude that

$$\theta_*^u(t; h, V_2) = q_*^{1-s} \cdot N_h(t) \tag{3.29}$$

for every $t \in \mathcal{O}_*$.

Equations (3.27), (3.28), and (3.29) tell us how to compute $\theta_*(t; h, \mathcal{O}^s)$. In order to collate the information they contain, we introduce the formal power series

$$G_h(\eta, z) \stackrel{\text{def}}{=} \sum_{\nu=0}^{\infty} \theta_*(\pi_*^\nu \eta; h, \mathcal{O}^s) \cdot (q_*^{-1} z)^\nu$$

in the indeterminate z, where η is a fixed element of $\mathbf{U}_*$. Our three equations imply that

$$\begin{aligned} G_h(\eta, z) &= \sum_{\nu=0}^{\infty} q_*^{d-s} \theta_*(\pi_*^{\nu-d}\eta; h, \mathcal{O}^s) \cdot (q_*^{-1} z)^\nu \\ &\quad + q_*^{1-s} \left[N_h(\eta) + \sum_{\nu=1}^{\infty} N_h(0) \cdot (q_*^{-1} z)^\nu \right] . \end{aligned}$$

After the change of index $\nu \leftarrow \nu + d$, the first summation on the right above equals $q_*^{-s} z^d \cdot G_h(\eta, z)$. The second term equals

$$q_*^{1-s} \left[N_h(\eta) + \frac{N_h(0) z}{q_* - z} \right] .$$

Therefore

$$(q_*^s - z^d) \cdot G_h(\eta, z) = q_* \left[N_h(\eta) + \frac{N_h(0) z}{q_* - z} \right]$$

which yields

$$G_h(\eta, z) = q_* \left[\frac{N_h(\eta)}{q_*^s - z^d} + \frac{N_h(0) z}{(q_* - z)(q_*^s - z^d)} \right] . \tag{3.30}$$

This identity shows that the values $\theta_*(\pi_*^\nu \eta; h, \mathcal{O}^s)$ can be computed for any $\nu \geq 0$ and any $\eta \in \mathbf{U}_*$ provided only that $N_h(0)$ and the finite set of cardinalities $\{N_h(\zeta) : \zeta \in \mathbf{K}(*)\}$ are known.

Theorem 3.14. *If $p \nmid d$ and $s > d$, then the derivative $\theta_*(t; h, \mathcal{O}^s)$ of the polynomial (3.25) exists as a continuous function on $\mathcal{O}_*$. Let $\delta = \mathbf{GCD}(d, q_* - 1)$. If $\delta < p$ and $s > \max\{1, \delta^2 - \delta\}$, then $\theta_*(t; h, \mathcal{O}^s)$ is bounded away from zero on $\mathcal{O}_*$.*

Proof. The existence and continuity of $\theta_*(t; h, \mathcal{O}^s)$ on $\mathcal{O}_*$ for $s > d$ is guaranteed by Theorem 3.12. For the second assertion, it suffices to show that $\theta_*(t; h, \mathcal{O}^s) \neq 0$ for any $t \in \mathcal{O}_*$ (as $\mathcal{O}_*$ is compact). For any $\zeta \in \mathbf{K}(*)$, the polynomial equation

$$\gamma_1 X_1^d + \cdots + \gamma_s X_s^d = \zeta$$

has a non-zero solution $(x_1, \ldots, x_s)$ in $\mathbf{K}(*)^s$ by Corollary 1.6. Therefore, $N_h(t)$ is never zero. By (3.30), this implies that the coefficients of the power series expansion of $(q_*^s - z^d) \cdot G_h(\eta, z)$ are all strictly positive. Therefore, $G_h(\eta, z)$ itself has strictly positive coefficients. We have now shown that $\theta_*(t; h, \mathcal{O}^s)$ is non-zero except possibly when $t = 0$. Putting $t = 0$ in (3.27) and substituting from (3.28) and (3.29) with $t = 0$, we find that

$$\theta_*(0; h, \mathcal{O}^s) = \frac{q_* N_h(0)}{q_*^s - q_*^d} .$$

Thus $\theta_*(t; h, \mathcal{O}^s)$ is never zero on $\mathcal{O}_*$. ■

Exercises 3.4

1. Assume $p \nmid d$. Show that $\theta_*(t; h, \mathcal{O}^s)$, with h given by (3.25), cannot exist as a continuous function on $\mathcal{O}_*$ for $s < d$.

2. Let $\theta : \mathcal{O}_* \longrightarrow \mathbf{C}$ be integrable for the measure d_*t. For $\eta \in \mathbf{U}_*$, let

$$G_\theta(\eta, z) = \sum_{\nu=0}^{\infty} \theta(\pi_*^\nu \eta) \cdot (q_*^{-1} z)^\nu .$$

Prove that $G_\theta(\eta, z)$ is absolutely convergent on the closed unit disk $|z| \leq 1$ for almost all $\eta \in \mathbf{U}_*$. Prove that

$$\int_{\mathbf{U}_*} G_\theta(\eta, 1)\, d_*\eta = \int_{\mathcal{O}_*} \theta(t)\, d_*t . \tag{3.31}$$

3. Check the evaluation (3.30) by integrating the righthand side at $z = 1$ over $\eta \in \mathbf{U}_*$. By (3.31), it should integrate out to one.

4. Take $k_* = \mathbf{Q}_p$ with h defined by (3.25). Relaxing the condition that $p \nmid d$, assume that $v_p(d) = \ell$. Write $\mu = \beta d + \rho$ with $0 \le \rho < d$, and put $N_h(\delta, t) = \#M_{t,V_2}(\delta)$ for all $\delta \ge 0$. Show that

$$\theta_p(p^\mu \eta; h) = p^{\delta(1-s)} \left[B \cdot N_h(\delta, 0) + p^{\beta(d-s)} \cdot N_h(\delta, p^\rho \eta) \right] \qquad (3.32)$$

for any fixed δ which is sufficiently large, where $B = 0$ if $\beta = 0$ but

$$\beta \ne 0 \Rightarrow B = \sum_{r=0}^{\beta-1} p^{r(d-s)} .$$

[Hint: It is convenient to prove this first with δ equal to a multiple of d.]

5. Show that (3.32) holds with $\delta = \phi$, where ϕ is the integer defined in Exercise 2.5(3). For $\gamma_1 = \gamma_2 = \ldots = \gamma_s = 1$, this is Theorem 2 of Hardy and Littlewood (1922b); but note the misprint in the definition of B.

6. Let $\mathbf{R}_+$ denote the positive real numbers. For real coefficients $\gamma_i > 0$, $1 \le i \le s$, let

$$h(x_1, x_2, \ldots, x_s) = \gamma_1 x_1^d + \gamma_2 x_2^d + \cdots + \gamma_s x_s^d .$$

Show that the Radon–Nikodym derivative $\theta(z)$ of the map $h : \mathbf{R}_+^s \to \mathbf{R}_+$ exists for all $d \ge 1$ and all integers $s \ge 1$. This means that $\theta(z)$ effects the change of variable

$$\int_{\mathbf{R}_+^s} f(h(x_1, x_2, \ldots, x_s))\, dx_1 dx_2 \cdots dx_s = \int_0^\infty f(z)\theta(z)\, dz \qquad (3.33)$$

for all those functions $f \in L^1(\mathbf{R}_+)$ for which the left hand side is defined. Show that $\theta(z)$ is homogeneous of weight $s/d - 1$. That is, show for $r > 0$ that

$$\theta(rz) = r^{\frac{s}{d}-1}\theta(z)$$

for almost all $z \in \mathbf{R}_+$. [Hint: Make the change of variables $x_i \leftarrow x_i / r^{1/d}$ on the left in (3.33).]

7. By choosing $f(z) = e^{-z}$ in (3.33), show that

$$\theta(z) = \frac{\Gamma(1 + 1/d)^s / \Gamma(s/d)}{(\gamma_1 \gamma_2 \cdots \gamma_s)^{1/d}} \cdot z^{\frac{s}{d}-1} .$$

4
The adèle ring over k

4.1 Definition and first properties

The *adèle ring* $\mathbb{A}_k$ over k is the restricted direct product over $\mathrm{v} \in \mathbf{M}$ of the completions k_v of k relative to their subrings $\mathcal{O}_\mathrm{v}$. This means that $\mathbb{A}_k$ is the subring of the product $\prod_{\mathrm{v}\in\mathbf{M}} k_\mathrm{v}$ which consists of the vectors $\mathbf{a} = (\mathbf{a}_\mathrm{v})_{\mathrm{v}\in\mathbf{M}}$ with $\mathbf{a}_\mathrm{v} \in \mathcal{O}_\mathrm{v}$ for almost all (i.e. all but finitely many) $\mathrm{v} \in \mathbf{M}$. The restricted direct product topology is defined on $\mathbb{A}_k$ by giving a base $\mathcal{B}$ for the neighbourhoods of zero:

$$B \in \mathcal{B} \iff B = \left(\prod_{\mathrm{v}\notin S} \mathcal{O}_\mathrm{v}\right) \times \left(\prod_{\mathrm{v}\in S} \pi_\mathrm{v}^{n_\mathrm{v}} \mathcal{O}_\mathrm{v}\right) \tag{4.1}$$

where S is any finite subset of $\mathbf{M}$ and where each n_v, for $\mathrm{v} \in S$, is a positive integer. The ring topology defined on $\mathbb{A}_k$ by $\mathcal{B}$ is locally compact because, e.g.,

$$D_0 = \left(\prod_{\mathrm{v}\in\mathbf{M}_{\mathrm{fin}}} \mathcal{O}_\mathrm{v}\right) \times \pi_\infty \mathcal{O}_\infty \in \mathcal{B} \tag{4.2}$$

is compact.

Each completion k_v is embedded as a subfield of $\mathbb{A}_k$ in a natural way via the inclusion map $i_\mathrm{v} : k_\mathrm{v} \longrightarrow \mathbb{A}_k$, which is defined in the following way: for $t_\mathrm{v} \in k_\mathrm{v}$, $i_\mathrm{v}(t_\mathrm{v})$ is the adèle which has v-component t_v and which has component zero at every other place of k. The ring $\mathbb{A}_k$ is therefore a global object containing the local information associated to the places of k. Further, each element $x \in k$ sits in $\mathbb{A}_k$ as the adèle which has component x at every $\mathrm{v} \in \mathbf{M}$. Therefore, k itself is embedded as a subfield of $\mathbb{A}_k$ 'along the diagonal'. The topology induced on k as a subfield of $\mathbb{A}_k$ is the discrete topology. To prove this fact, it suffices to show that $k \cap D_0 = \{0\}$. Now any element of k lying in D_0 has no pole and must therefore be a constant. Since the only constant function in $\pi_\infty \mathcal{O}_\infty$ is zero, $k \cap D_0 = \{0\}$.

In our work, the ∞-component of $\mathbb{A}_k$ plays a special role because T is the generator of k which we have chosen as the indeterminate to use in writing polynomials over $\mathbb{F}_q$. Identifying k_∞ with its component $i_\infty(k_\infty)$ in $\mathbb{A}_k$, we write

$$\mathbb{A}_k = \tilde{\mathbb{A}}_k \times k_\infty \tag{4.3}$$

where $\tilde{\mathbb{A}}_k$ is the ideal in $\mathbb{A}_k$ consisting of the adèles with ∞-component equal to zero. We call $\tilde{\mathbb{A}}_k$ the ring of *finite* adèles, and for $t \in \mathbb{A}_k$ we write

$$t = \tilde{t} + t_\infty$$

where $\tilde{t} \in \tilde{\mathbb{A}}_k$ and where t_∞ is the component of t at ∞.

The adèle ring of any global field can be defined as a restricted direct product over its places (see Cassels 1967 or Lang 1970). For the rational function field k, the structure of the adèle ring is especially transparent.

Proposition 4.1. *The topological ring $\mathbb{A}_k$ is the direct product of k (embedded as a discrete subfield along the diagonal) and the compact subring D_0 defined by* (4.2).

Proof. For a given adèle $\mathbf{a} = (\mathbf{a}_v)_{v \in \mathbf{M}}$, let $S(\mathbf{a})$ be the finite set of places $v \in \mathbf{M}_{\text{fin}}$ for which $\mathbf{a}_v \notin \mathcal{O}_v$. For $P \in S(\mathbf{a})$, define the *principal part* of $\mathbf{a}$ at P to be the sum of the terms in the P-adic expansion (2.4) of $\mathbf{a}_P$ which have negative exponents. Since each such principal part lies in k, their sum x is also in k; and $\mathbf{a} - x$ is an adèle which has integral v-component at every finite place of k. The sum of the terms in the Laurent series expansion (2.3) of $(\mathbf{a} - x)_\infty$ which have negative or zero exponent is a polynomial A in $\mathbf{A}$ and is therefore integral at every finite place. Thus, $\mathbf{a} - x - A$ is an element of D_0; and $\mathbf{a} = (A + x) + (\mathbf{a} - x - A)$ is a decomposition of $\mathbf{a}$ with summands in k and D_0 respectively. Since $k \cap D_0 = \{0\}$, this decomposition is unique. ∎

Corollary 4.2. *The quotient $\mathbb{A}_k/k$ is a compact additive group which is isomorphic to D_0.*

Proof. Since D_0 is an open subgroup of $\mathbb{A}_k$, the restriction to D_0 of the canonical map onto the quotient is bi-continuous. ∎

Definition 4.3. *The group $\mathbb{A}_k/k$ is called the* adèle class group, *and its elements are called* adèle classes. *Let* **can** : $\mathbb{A}_k \longrightarrow \mathbb{A}_k/k$ *be the canonical map. A compact open subset D of $\mathbb{A}_k$ for which the restriction* **can** : $D \longrightarrow \mathbb{A}_k/k$ *is a bijection is called a* fundamental domain *for $\mathbb{A}_k/k$.*

For example, D_0 is such a fundamental domain. Of course there are many other fundamental domains.

In the classical Hardy–Littlewood 'circle method', partition problems in $\mathbb{Z}$ are solved by integrating appropriate generating functions over the unit circle in the complex plane. For the study of partition problems in $\mathbf{A}$, the most convenient analogue for the circle is the adèle class group. The complex path integral used by Hardy–Littlewood is replaced by the Haar integral on $\mathbb{A}_k/k$.

As we note in §4.3, the character group of $\mathbb{A}_k/k$ is the discrete group k^+. Since $\mathbb{A}_k/k$ is also the character group of k^+ (Pontryagin duality), the adèle class group as an invariant of the field k can be defined in a very natural way: $\mathbb{A}_k/k = \hat{k}^+$.

4.2 Adèlic harmonic analysis

For $\mathrm{v} \in \mathbf{M}_{\mathrm{fin}}$, let $d_{\mathrm{v}}t_{\mathrm{v}}$ be the Haar measure on k_{v} which gives $\mathcal{O}_{\mathrm{v}}$ measure one; and let $d_\infty t_\infty$ be the Haar measure on k_∞ which gives $\pi_\infty \mathcal{O}_\infty$ measure one. These measures induce a Haar measure on $\mathbb{A}_k$ as follows: let $f_{\mathrm{v}} \in L^1(k_{\mathrm{v}})$ be any sequence of functions with the property that f_{v} restricts to the constant function 1 on $\mathcal{O}_{\mathrm{v}}$ for almost all $\mathrm{v} \in \mathbf{M}$. Define $f(t)$ for $t = (t_{\mathrm{v}})_{\mathrm{v}\in\mathbf{M}} \in \mathbb{A}_k$ by

$$f(t) = \prod_{\mathrm{v}\in\mathbf{M}} f_{\mathrm{v}}(t_{\mathrm{v}}) \,, \tag{4.4}$$

a finite product since $t_{\mathrm{v}} \in \mathcal{O}_{\mathrm{v}}$ for almost all v. We say that $f(t)$ is *integrable* if

$$\prod_{\mathrm{v}\in\mathbf{M}} \int_{k_{\mathrm{v}}} |f_{\mathrm{v}}(t)|\, dt_{\mathrm{v}} < \infty \,. \tag{4.5}$$

The integral of f over $\mathbb{A}_k$ is then defined by

$$\int_{\mathbb{A}_k} f(t)\, dt = \prod_{\mathrm{v}\in\mathbf{M}} \int_{k_{\mathrm{v}}} f_{\mathrm{v}}(t_{\mathrm{v}})\, dt_{\mathrm{v}} \,. \tag{4.6}$$

This integral is clearly invariant under translation by elements of $\mathbb{A}_k$. One can show that the linear span of the functions f of type (4.4) which satisfy (4.5) is rich enough to define a Haar measure dt on $\mathbb{A}_k$ by the usual measure theoretic extension of the integral (4.6) (cf. Lang 1970).

Let 'dt' also denote the projection of the measure dt onto the adèle class group $\mathbb{A}_k/k$. If D is any fundamental domain for $\mathbb{A}_k/k$, then by definition

$$\int_{\mathbb{A}_k/k} f(t)\, dt = \int_D f(\mathbf{can}(t))\, dt$$

for all $f \in L^1(\mathbb{A}_k/k)$. Since the characteristic function of D_0 is of type (4.4), the dt-measure of D_0 is easily computed and is found to be unity.

Therefore, the projected measure is a Haar measure on $\mathbb{A}_k/k$ of total mass one.

Having described Haar measures on $\mathbb{A}_k$ and $\mathbb{A}_k/k$, we turn now to a description of their character groups. We note first that if $\Lambda \in \widehat{\mathbb{A}}_k$, then $\Lambda_v = \Lambda \circ i_v \in \hat{k}_v$ for all $v \in \mathbf{M}$.

Proposition 4.4. *Let $\Lambda \in \widehat{\mathbb{A}}_k$. Then for almost all $v \in \mathbf{M}$, the local conductor $n(\Lambda_v) \leq 0$.*

Proof. Let $\mathcal{H}$ be the open half-plane in $\mathbb{C}$ consisting of the points with real part greater than zero. Then $\Lambda^{-1}(\mathcal{H})$ is an open neighbourhood of zero in $\mathbb{A}_k$ and therefore contains a basic open neighbourhood B of type (4.1). Now B is a subgroup of the additive group of $\mathbb{A}_k$ and therefore maps under Λ to a subgroup of the part of the unit circle lying in $\mathcal{H}$. Since there is no subgroup of the circle other than $\{1\}$ lying in $\mathcal{H}$, $\Lambda(B) = 1$ and the proposition is proved. ■

This proposition shows that Λ can be decomposed into a product of local factors. In fact since Λ is continuous,

$$\Lambda(t) = \prod_{v \in \mathbf{M}} \Lambda_v(t_v) \tag{4.7}$$

for all $t = (t_v)_{v \in \mathbf{M}}$, the product on the right being actually a finite product by the proposition. Conversely if $\Lambda_v \in \hat{k}_v$ is a sequence of local characters such that $n(\Lambda_v) \leq 0$ for almost all $v \in \mathbf{M}$, then (4.7) defines a character $\Lambda \in \widehat{\mathbb{A}}_k$.

Definition 4.5. *For $\Lambda \in \widehat{\mathbb{A}}_k$ and $\mathbf{a} \in \mathbb{A}_k$, we define $\Lambda^{\mathbf{a}} \in \widehat{\mathbb{A}}_k$ by*

$$\Lambda^{\mathbf{a}}(t) = \Lambda(\mathbf{a}t)$$

for all $t \in \mathbb{A}_k$ (cf. (2.11)).

Proposition 4.6. *Let $\Gamma \in \widehat{\mathbb{A}}_k$ be a fixed character having the property that $n(\Gamma_v) = 0$ for almost all $v \in \mathbf{M}$ and $n(\Gamma_v) > -\infty$ for all $v \in \mathbf{M}$. Then the mapping $\phi_\Gamma : \mathbb{A}_k \longrightarrow \widehat{\mathbb{A}}_k$ defined by $\phi_\Gamma(\mathbf{a}) = \Gamma^{\mathbf{a}}$ is an algebraic and topological isomorphism from $\mathbb{A}_k$ onto $\widehat{\mathbb{A}}_k$.*

Proof. Let Λ be any character in $\widehat{\mathbb{A}}_k$. By Proposition 2.1, for any $v \in \mathbf{M}$ there is an element $\mathbf{a}_v \in k_v$ such that $\Lambda_v(t_v) = \Gamma_v(\mathbf{a}_v t_v)$ for all $t_v \in k_v$. Now by (2.12)

$$v(\mathbf{a}_v) = n(\Gamma) - n(\Lambda) \geq -n(\Lambda)$$

for almost all $v \in \mathbf{M}$. Therefore the vector $\mathbf{a} = (\mathbf{a}_v)_{v \in \mathbf{M}}$ is an adèle and then (4.7) shows that $\Lambda = \Gamma^{\mathbf{a}}$. We have now shown that ϕ_Γ is onto. The fact that ϕ_Γ is bi-continuous is a straightforward consequence of the definitions and Proposition 2.1. ■

4.3 The character E

In this section, we define the character $E \in \widehat{\mathbb{A}}_k$ which will be used in Chapters 6 and 8 to create generating functions for the polynomial 3-primes problem and the Waring problem. Since we define E in such a way that $E(k) = 1$, E projects to a character of the adèle class group $\mathbb{A}_k/k$.

Let $e \in \widehat{\mathbb{F}}_q$ be the standard character defined for $\alpha \in \mathbb{F}_q$ by

$$e(\alpha) = e^{2\pi i \cdot \mathrm{Tr}(\alpha)/p} \tag{4.8}$$

where $\mathrm{Tr}(\alpha) \in \mathbb{F}_p$ is the absolute trace of α. We first define $E_\infty \in \hat{k}_\infty$ as follows: for $x \in k_\infty$

$$E_\infty(x) = e(\alpha_1) \tag{4.9}$$

where α_1 is the coefficient of $1/T$ in the Laurent expansion (2.3) of x.

Definition 4.7. *By Proposition* 4.1, *any $t \in \mathbb{A}_k$ can be decomposed in a unique way as the sum of an element $\delta \in D_0$ and an element $z \in k$: $t = \delta + z$. We define $E \in \widehat{\mathbb{A}}_k$ by*

$$E(t) = E_\infty(\delta_\infty)$$

for all $t \in \mathbb{A}_k$.

Proposition 4.8. *The character E satisfies:* (1) $E(k) = 1$. (2) $n(E_v) = 0$ for all $v \in \mathbf{M}_{\mathrm{fin}}$. (3) $n(E_\infty) = 2$.

Proof. Properties (1) and (3) are obvious from the definitions. For (2), note first from Definition 4.7 that $n(E_v) \leq 0$ at every place $v \in \mathbf{M}_{\mathrm{fin}}$. Fix a positive irreducible polynomial P of degree d, and let 'P' also denote the finite place associated to P. Put $z = T^{d-1}/P \in k$. Since z is integral at every place $v \neq P$ or ∞,

$$1 = E(\beta z) = E_P(\beta z)E_\infty(\beta z) \tag{4.10}$$

for all $\beta \in \mathbb{F}_q$. Now the first term in the Laurent expansion (2.3) of $1/P$ is $(1/T)^d$. Therefore

$$E_\infty(\beta z) = e^{2\pi i \cdot \mathrm{Tr}(\beta)/p} \neq 1$$

if we choose β so that $\mathrm{Tr}(\beta) \neq 0$. It follows then from (4.10) that $E_P(\beta z) \neq 1$. Thus $n(E_P) \geq 0$. ∎

Since $n(E_v) = 0$ for all $v \in \mathbf{M}_{\mathrm{fin}}$, we can set up an isomorphism between $\mathbb{A}_k$ and $\widehat{\mathbb{A}}_k$ by choosing $\Gamma = E$ in Proposition 4.6. This choice leads to a statement of the *Fourier inversion theorem* for $\mathbb{A}_k$ which is most convenient for our work. For $f \in L^1(\mathbb{A}_k)$, define the Fourier transform $\hat{f}$ of f as follows:

$$\hat{f}(x) = \int_{\mathbb{A}_k} f(t)E(xt)\,dt$$

for all $x \in \mathbb{A}_k$. Then we have

Proposition 4.9. (Fourier Inversion in $\mathbb{A}_k$) *If $f \in L^1(\mathbb{A}_k)$ is continuous and if $\hat{f} \in L^1(\mathbb{A}_k)$, then*

$$f(t) = q^2 \int_{\mathbb{A}_k} \hat{f}(x)E(-xt)\,dx \tag{4.11}$$

for all $t \in \mathbb{A}_k$.

Proof. From the general theory, we know that (4.11) holds for some constant multiplier of the integral on the right; and we must show that this multiplier is q^2. Let f be any positive integrable non-negligible function of type (4.4). From the definitions,

$$\hat{f}(x) = \prod_{\mathrm{v}\in\mathbf{M}} \hat{f}_{\mathrm{v}}(x_{\mathrm{v}})$$

so that $\hat{f}$ is also of type (4.4). One can choose f so that $\hat{f}$ is integrable (e.g. let f be the characteristic function of D_0). Then by Proposition 2.3,

$$\begin{aligned}\int_{\mathbb{A}_k} \hat{f}(x)E(-xt)\,dx &= \prod_{\mathrm{v}\in\mathbf{M}} \int_{k_{\mathrm{v}}} \hat{f}_{\mathrm{v}}(x_{\mathrm{v}})E(-x_{\mathrm{v}}t_{\mathrm{v}})\,dt_{\mathrm{v}} \\ &= \prod_{\mathrm{v}\in\mathbf{M}} q^{-n(E_{\mathrm{v}})} f_{\mathrm{v}}(t_{\mathrm{v}}) \\ &= q^{-2} f(t)\end{aligned}$$

since $n(E_{\mathrm{v}}) = 0$ for $\mathrm{v} \in \mathbf{M}_{\mathrm{fin}}$ and $n(E_\infty) = 2$. ∎

We will also need a Fourier inversion theorem for the group of finite adèles $\tilde{\mathbb{A}}_k$. From the direct sum decomposition (4.3), we have $dt = d\tilde{t}\,dt_\infty$, where $d\tilde{t}$ is the projection of the measure dt onto $\tilde{\mathbb{A}}_k$. For $f \in L^1(\tilde{\mathbb{A}}_k)$, define the Fourier transform $\hat{f}$ of f as follows:

$$\hat{f}(\tilde{x}) = \int_{\tilde{\mathbb{A}}_k} f(\tilde{t})E(\tilde{x}\tilde{t})\,d\tilde{t}$$

for all $\tilde{x} \in \tilde{\mathbb{A}}_k$. Then we have

Proposition 4.10. (Fourier Inversion in $\tilde{\mathbb{A}}_k$) *If $f \in L^1(\tilde{\mathbb{A}}_k)$ is continuous and if $\hat{f} \in L^1(\tilde{\mathbb{A}}_k)$, then*

$$f(\tilde{t}) = \int_{\tilde{\mathbb{A}}_k} \hat{f}(\tilde{x})E(-\tilde{x}\tilde{t})\,d\tilde{x}$$

for all $\tilde{t} \in \tilde{\mathbb{A}}_k$.

Proof. The proof goes the same way as the proof of Proposition 4.9 except that there is no contribution from the infinite place. ■

The character E is also useful in computing the character group of $\mathbb{A}_k/k$. We know from the general theory that this character group is isomorphic to

$$k^{\perp} = \{\mathbf{a} \in \mathbb{A}_k : E(\mathbf{a}t) = 1 \text{ for all } t \in k\}\,.$$

Proposition 4.11. *The character group of $\mathbb{A}_k/k$ is isomorphic to the discrete group k^+.*

Proof. Since $E(k) = 1$, $k \subseteq k^{\perp}$. Conversely, suppose $\mathbf{a} = \delta + z \in k^{\perp}$ with $\delta \in D_0$ and $z \in k$ (cf. Proposition 4.1). Then

$$1 = E(\mathbf{a}t) = E(\delta t)E(zt) = E(\delta t)$$

for all $t \in k$. We will show that $\delta = 0$. First choose $t = A \in \mathbf{A}$. Then

$$1 = E(\delta t) = E_{\infty}(\delta_{\infty} A) \tag{4.12}$$

because $\delta A \in \mathcal{O}_{\mathrm{v}}$ for every finite place v. Since (4.12) holds in particular for $A = \beta T^d$ with $\beta \in \mathbb{F}_q$ and $d \geq 0$, we conclude from the definition (4.9) that all the coefficients in the Laurent expansion (2.3) of δ_{∞} are zero. Thus $\delta_{\infty} = 0$.

Next choose $t = A/P^{\nu}$ where P is a positive irreducible, $\nu \in \mathbb{Z}$ with $\nu \geq 0$, and $A \in \mathbf{A}$. Then

$$1 = E(\delta t) = E_P(\delta_P A/P^{\nu}) \tag{4.13}$$

because $\delta A/P^{\nu} \in \mathcal{O}_{\mathrm{v}}$ for every finite place $\mathrm{v} \neq P$ while $\delta_{\infty} = 0$. The rational functions A/P^{ν} contain a complete set of representatives for $k_P/\mathcal{O}_P$, which is the character group of $\mathcal{O}_P$ by Proposition 2.2. Therefore (4.13) shows that every character in $\widehat{\mathcal{O}}_P$ is trivial on δ_P, and this proves that $\delta_P = 0$. Now $\delta = 0$ because $\delta_{\mathrm{v}} = 0$ for every $\mathrm{v} \in \mathbf{M}$. ■

Exercises 4.3

1. Does E restrict to E_∞ on $i_\infty(k_\infty)$?

2. Prove that the character E can be defined using the theory of residues (Serre 1988) as follows: let $\mathbf{P}^1$ be the projective line over $\mathbf{F}_q$. For $\mathbf{a} \in \mathbb{A}_k$, the differential $\mathbf{a}_v dT$ is welldefined at every point $v \in \mathbf{P}^1$. Let $\mathrm{res}_v(\mathbf{a}_v dT) \in \mathbf{F}_q$ be the sum of the residues of $\mathbf{a}_v dT$ at all geometric points over v. Then

$$E(\mathbf{a}) = e\left(\sum_{v \in \mathbf{M}_{\mathrm{fin}}} \mathrm{res}_v(\mathbf{a}_v dT)\right).$$

4.4 Global Radon–Nikodym derivatives

The ring of finite adèles $\tilde{\mathbb{A}}_k$ contains the compact open subring

$$\mathcal{O} \stackrel{\mathrm{def}}{=} \prod_{P \in \mathbf{M}_{\mathrm{fin}}} \mathcal{O}_P, \tag{4.14}$$

and the Haar measure $d\tilde{t}$ on $\tilde{\mathbb{A}}_k$ restricts to the product measure $\prod_P d_P t_P$ on $\mathcal{O}$. The ring $\mathcal{O}$ therefore has unit total mass in the measure $d\tilde{t}$.

In this section, we will globalize the theory of local Radon–Nikodym derivatives developed in Chapters 2 and 3 to the ring $\mathcal{O}$. For every $P \in \mathbf{M}_{\mathrm{fin}}$, let V_P be a given compact open subset of $\mathcal{O}_P^s$, and put

$$V = \prod_{P \in \mathbf{M}_{\mathrm{fin}}} V_P,$$

a compact but not necessarily open subset of $\mathcal{O}^s$. Let $\rho_{V,P}$ be the restriction of the product measure on $\mathcal{O}_P^s$ to V_P, normalized so that V_P gets unit total mass, and put $\rho_V = \prod_P \rho_{V,P}$. Any polynomial

$$h(X_1, ..., X_s) \in \mathbf{A}[X_1, \ldots, X_s]$$

defines a map $h_P : V_P \to \mathcal{O}_P$ for every $P \in \mathbf{M}_{\mathrm{fin}}$ because the coefficients of h are integral at every finite place of k. We assume throughout this section that the local Radon–Nikodym derivatives $\theta_P(t_P; h_P, V_P)$ as defined in §2.4 exist as continuous functions on $\mathcal{O}_P$ at every $P \in \mathbf{M}_{\mathrm{fin}}$.

Theorem 4.12. *Assume that there is a sequence of positive real numbers $c(P)$ with $\sum_P c(P) < \infty$ such that*

$$\sup_{t_P \in \mathcal{O}_P} |\theta_P(t_P; h_P, V_P) - 1| \le c(P) \tag{4.15}$$

for all $P \in \mathbf{M}_{\text{fin}}$. Then the product

$$\theta(\tilde{t}; h, V) \stackrel{\text{def}}{=} \prod_{P \in \mathbf{M}_{\text{fin}}} \theta_P(t_P; h_P, V_P) \tag{4.16}$$

converges uniformly for $\tilde{t} = (t_P) \in \mathcal{O}$ to a continuous function on $\mathcal{O}$. Further, $\theta(\tilde{t}; h, V)$ is the Radon–Nikodym derivative of the map induced by h from the measure space (V, ρ_V) to the measure space $(\mathcal{O}, d\tilde{t})$.

Proof. By (4.15), the set

$$S = \{P \in \mathbf{M}_{\text{fin}} : \theta_P(t_P; h_P, V_P) \text{ is bounded away from zero on } \mathcal{O}_P\}$$

has finite complement in $\mathbf{M}_{\text{fin}}$. Now (4.15) also implies that

$$\sum_{P \in S} \log \theta_P(t_P; h_P, V_P)$$

is a uniformly convergent series of continuous functions on $\mathcal{O}$. The product over $P \in S$ in (4.16) is therefore continuous. Since the product over $P \notin S$ is finite, we see that $\theta(\tilde{t}; h, V)$ is itself continuous.

For a given finite subset I of $\mathbf{M}_{\text{fin}}$, let Γ_I be the set of functions $f \in C(\mathcal{O})$ of the form

$$f(\tilde{t}) = \prod_{P \in \mathbf{M}_{\text{fin}}} f_P(t_P),$$

where each $f_P \in C(\mathcal{O}_P)$ and where $f_P = 1$ for $P \notin I$. For $f \in \Gamma_I$,

$$f(\tilde{t}) \cdot \theta(\tilde{t}; h, V) = \lim_J \prod_{P \in J} f_P(t_P) \cdot \theta_P(t_P; h_P, V_P) \tag{4.17}$$

in the uniform topology on $C(\mathcal{O})$, where J runs through the finite subsets of $\mathbf{M}_{\text{fin}}$ containing I. Put $\tilde{\mathbf{t}} = (\tilde{t}^{(1)}, \ldots, \tilde{t}^{(s)}) \in \mathcal{O}^s$ and $\mathbf{t}_P = (t_P^{(1)}, \ldots, t_P^{(s)}) \in \mathcal{O}_P^s$. Then by definition of the product measure, for $f \in \Gamma_I$ and for any finite $J \supseteq I$ we have

$$\int_V f(h(\tilde{\mathbf{t}}))\, d\rho_V(\tilde{\mathbf{t}}) \quad = \quad \prod_{P \in J} \int_{V_P} f_P(h_P(\mathbf{t}_P))\, \frac{d\mathbf{t}_P}{meas(V_P)}$$

$$= \prod_{P\in J}\int_{\mathcal{O}_P} f(t_P)\cdot\theta_P(t_P;h_P,V_P)\,dt_P$$

$$= \int_{\mathcal{O}_J}\prod_{P\in J} f(t_P)\cdot\theta_P(t_P;h_P,V_P)\,dt_P$$

where $\mathcal{O}_J = \prod_{P\in J}\mathcal{O}_P$. Taking the limit over J and appealing to (4.17), we get

$$\int_V f(h(\tilde{\mathbf{t}}))\,d\rho_V(\tilde{\mathbf{t}}) = \int_{\mathcal{O}} f(\tilde{t})\cdot\theta(\tilde{t};h,V)\,d\tilde{t}. \tag{4.18}$$

Since the linear span of the functions in $\cup_I\Gamma_I$ is dense in $L^1(\mathcal{O})$, the identity (4.18) holds in fact for all $f\in L^1(\mathcal{O})$. ∎

By Proposition 2.2 and a general theorem of harmonic analysis,

$$\widehat{\mathcal{O}} \simeq \coprod_P \widehat{\mathcal{O}}_P \simeq \coprod_P k_P/\mathcal{O}_P, \tag{4.19}$$

the notation '$\coprod$' indicating the direct sum. Using the global character E, we can define a continuous pairing $\mathcal{O}\times(k/\mathbf{A}) \to \mathbb{C}$ which induces the isomorphisms of (4.19) in explicit form.

Definition 4.13. *Let A and Q be polynomials in $\mathbf{A}$ with $(A,Q)=1$ and Q positive. For $\tilde{t}=(t_P)\in\mathcal{O}\subset\tilde{\mathbb{A}}_k$, we define*

$$\langle\tilde{t},A/Q\rangle \overset{\text{def}}{=} E(\tilde{t}A/Q) = \prod_P E_P(t_PA/Q). \tag{4.20}$$

Theorem 4.14. *The pairing* (4.20) *has zero kernel in its first argument; and the kernel in its second argument is the polynomial ring* $\mathbf{A}$. *Therefore the pairing induces isomorphisms*

$$\widehat{\mathcal{O}} \simeq k/\mathbf{A} \simeq \coprod_{P\in\mathbf{M}_{\text{fin}}} k_P/\mathcal{O}_P. \tag{4.21}$$

Proof. Suppose first that $\langle\tilde{t},A/Q\rangle = E(\tilde{t}A/Q)=1$ for all rational functions $A/Q\in k$. For fixed $P\in\mathbf{M}_{\text{fin}}$, it follows from (2) of Proposition 4.8 that $E_P(t_PA/P^\nu) = E(\tilde{t}A/P^\nu)=1$ for all A prime to P and all $\nu\geq 0$. Therefore $t_P=0$ by Proposition 2.2. This being true for all P, we conclude that $\tilde{t}=0$.

Suppose next that $\langle\tilde{t},A/Q\rangle = E(\tilde{t}A/Q)=1$ for all $\tilde{t}\in\mathcal{O}$. Taking $\tilde{t}=i_P(t_P)$, we see that $E_P(t_PA/Q)=1$ for all $t_P\in\mathcal{O}_P$. Now (2) of Proposition 4.8 implies that $E_P(t_PA/Q) = E_P(t_PB/P^\nu)$ where B/P^ν

with $B \in \mathbf{A}$ is the P-part of the partial fraction decomposition of A/Q in k. Since t_P is an arbitrary element of $\mathcal{O}_P$, Proposition 2.2 implies that $B/P^\nu \equiv 0 \pmod{\mathbf{A}}$. This being true for all P, we conclude that $A/Q \in \mathbf{A}$. ∎

As is clear from the proof of this theorem, the second isomorphism in (4.21) is induced by the partial fraction decomposition in k.

Let $\mathcal{R}$ be the set of rational functions $A/Q \in k$ such that: (1) $Q \in \mathbf{A}$ is positive, and (2) A is a polynomial in $\mathbf{A}$ which is prime to Q and satisfies $\deg A < \deg Q$. The rational functions A/Q of this type constitute a complete set of representatives for $k/\mathbf{A}$. The pairing (4.20) and the isomorphisms (4.21) enable us to formulate the Fourier transform of a function $f \in \mathrm{L}^1(\mathcal{O})$ as follows: for $A/Q \in \mathcal{R}$,

$$\hat{f}(A/Q) = \int_{\mathcal{O}} f(\tilde{t}) \cdot E(A\tilde{t}/Q)\, d\tilde{t}. \tag{4.22}$$

For the Radon–Nikodym derivative $\theta(\tilde{t}; h, V)$, we have

$$\begin{aligned}\hat{\theta}(A/Q; h, V) &= \int_{\mathcal{O}} \theta(\tilde{t}; h, V) \cdot E(A\tilde{t}/Q)\, d\tilde{t} \\ &= \prod_{P \in \mathbf{M}_{\mathrm{fin}}} \int_{\mathcal{O}_P} \theta_P(t_P; h_P, V_P) \cdot E_P(At_P/Q)\, dt_P \qquad (4.23) \\ &= \prod_{P \in \mathbf{M}_{\mathrm{fin}}} \hat{\theta}_P(A/Q; h_P, V_P)\end{aligned}$$

by (4.16). Thus each global Fourier transform of the Radon–Nikodym derivative of h is the product of its local Fourier transforms.

Whenever a summation over A/Q appears subsequently in the text, it is understood that that summation is over the rational functions in $\mathcal{R}$. Thus the formal Fourier series associated to a function $f \in \mathrm{L}^1(\mathcal{O})$ evaluated at $\tilde{t} \in \mathcal{O}$ may be written as

$$\sum_{A/Q} \hat{f}(A/Q) \cdot E(-A\tilde{t}/Q). \tag{4.24}$$

4.4.1 The derivative of $X_1 + X_2 + \cdots + X_s$ on $V = \prod_P \mathbf{U}_P^s$

By (3.12), the local derivative of the linear polynomial

$$h(X_1, \ldots, X_s) = X_1 + \ldots + X_s \tag{4.25}$$

on $V_P = \mathbf{U}_P$ is given by

$$\theta_P(t_P; h_P, V_P) = \begin{cases} 1 + \frac{(-1)^{s-1}}{(|P|-1)^s} & \text{if } \mathrm{v}_P(t_P) = 0 \\ 1 - \frac{(-1)^{s-1}}{(|P|-1)^{s-1}} & \text{if } \mathrm{v}_P(t_P) > 0, \end{cases} \tag{4.26}$$

a continuous function on $\mathcal{O}_P$, where $|P|$ is the absolute value (2.2) of P at ∞. Since the number of positive irreducibles P of degree r is bounded above by q^r,

$$\sum_{P \in \mathbf{M}_{\text{fin}}} |\theta_P(t_P; h_P, V_P) - 1| \leq \sum_{r=1}^{\infty} q^r (q^r - 1)^{1-s}$$

which converges for $s \geq 3$. Invoking Theorem 4.12, we conclude the following.

Theorem 4.15. *The global Radon–Nikodym derivative of* (4.25) *on* $V = \prod_P \mathbf{U}_P^s$ *exists as a continuous function on* $\mathcal{O}$. *For* $s = 3$,

$$\theta(\tilde{t}; h, V) = \prod_{\substack{P \\ \mathrm{v}_P(t_P)=0}} \left(1 + \frac{1}{(|P|-1)^3}\right) \prod_{\substack{P \\ \mathrm{v}_P(t_P)>0}} \left(1 - \frac{1}{(|P|-1)^2}\right). \tag{4.27}$$

For $s \geq 3$, we may expand the derivative $\theta(\tilde{t}; h, V)$ in its Fourier series. For this purpose, it is convenient to introduce the polynomial analogues of the Möbius μ-function and the Euler ϕ-function. For a positive $Q \in \mathbf{A}$, let

$$Q = \prod_{i=1}^{a} P_i^{\nu_i} \tag{4.28}$$

be the prime decomposition of Q. Then

$$\mu(Q) \stackrel{\text{def}}{=} \begin{cases} 0 & \text{if any } \nu_i \geq 2 \\ (-1)^a & \text{otherwise} \end{cases} \tag{4.29}$$

and

$$\Phi(Q) \stackrel{\text{def}}{=} |Q| \cdot \prod_{i=0}^{a} \left(1 - \frac{1}{|P_i|}\right). \tag{4.30}$$

Of course, $\Phi(Q)$ is the order of the multiplicative group of the quotient ring $\mathbf{A}/Q\mathbf{A}$.

We may now rewrite (3.10) at $P \in \mathbf{M}_{\text{fin}}$ as

$$\hat{\theta}_P(A/Q; h_P, V_P) = \frac{\mu(P^\nu)^s}{\Phi(P^\nu)^s}$$

where P^ν is the largest power of P dividing Q. It follows from this last equation and (4.23) that

$$\hat{\theta}(A/Q; h, V) = \prod_{P \in \mathbf{M}_{\text{fin}}} \frac{\mu(P^\nu)^s}{\Phi(P^\nu)^s} = \frac{\mu(Q)^s}{\Phi(Q)^s} \tag{4.31}$$

so that formally we have

$$\theta(\tilde{t}; h, V) = \sum_{A/Q} \frac{\mu(Q)^s}{\Phi(Q)^s} \cdot E(-A\tilde{t}/Q). \tag{4.32}$$

The absolute convergence of the above series for $s \geq 3$ will follow after we prove two lemmas.

Lemma 4.16. *For all positive* $Q \in \mathbf{A}$, *we have*

$$\Phi(Q) \geq \begin{cases} \sqrt{|Q|} & \text{if } Q \text{ is odd} \\ \frac{1}{2}\sqrt{|Q|} & \text{if } Q \text{ is even.} \end{cases} \tag{4.33}$$

Proof. Assume first that Q is odd. Since $\Phi(Q)$ is a multiplicative function of Q, there is no loss of generality in assuming that $Q = P^\nu$ is a prime power. If $\nu \geq 2$, then

$$\Phi(Q)/\sqrt{|Q|} = |P|^{\frac{\nu}{2}-1}(|P| - 1) \geq 1$$

while if $\nu = 1$ but $|P| \geq 3$, then

$$\Phi(Q)/\sqrt{|Q|} = |P|^{\frac{1}{2}} - |P|^{-\frac{1}{2}} \geq 1$$

as required. This proves (4.33) for odd Q.

For even Q, the above estimates hold for all the maximal prime power divisors of Q except possibly for the divisors $P = T$ or $P = T+1$. For such a divisor, $\Phi(P)/\sqrt{|P|} = 1/\sqrt{2}$. If both T and $T+1$ divide Q to the first power, which is the worst case, we get the second estimate in (4.33). ∎

The lower bound (4.33) implies the absolute convergence of the right hand side of (4.32) for any $s \geq 4$. To establish the absolute convergence for $s = 3$, we can use the following bound.

Lemma 4.17. *For every degree $j \geq 1$, we have*

$$\sum_{\deg Q = j}{}' \frac{1}{\Phi(Q)} \leq \begin{cases} \frac{3}{4}(j+1) & \text{if } q > 2 \\ j+1 & \text{if } q = 2, \end{cases} \tag{4.34}$$

the prime on the summation sign indicating a sum over positive polynomials only.

Proof. Suppose Q has prime decomposition (4.28) and degree j. Then

$$\Phi(Q) = |Q| \cdot \prod_{i=1}^{a} \left(1 - \frac{1}{|P_i|}\right) \geq q^j \left(\frac{1}{2}\right)^a. \tag{4.35}$$

Therefore, $1/\Phi(Q) \leq q^{-j}\tau(Q)$, where $\tau(Q)$ is the number of positive divisors of Q in $\mathbf{A}$. It follows that

$$\sum_{\deg Q = j}{}' \frac{1}{\Phi(Q)} \leq q^{-j} \sum_{\deg Q = j}{}' \tau(Q) = j + 1,$$

because Carlitz (1932) has shown that the average value of the divisor function τ on the positive polynomials of degree j is just $j+1$ (see also Exercise 5.4(1)). This proves (4.34) when $q = 2$.

In case $q \geq 3$, (4.35) can be improved to $\Phi(Q) \geq (4/3)q^j(1/2)^a$. The proof then proceeds as before. ∎

Now since there are exactly $\Phi(Q)$ fractions A/Q with fixed denominator Q in the summation of (4.32), we have

$$\begin{aligned} \sum_{A/Q} \left| \frac{\mu(Q)^s}{\Phi(Q)^s} \cdot E(-A\tilde{t}/Q) \right| &\leq \sum_{j=0}^{\infty} \sum_{\deg Q = j}{}' \frac{1}{\Phi(Q)^{s-1}} \\ &\leq 2 \sum_{j=0}^{\infty} |Q|^{\frac{2-s}{2}} \sum_{\deg Q = j}{}' \frac{1}{\Phi(Q)} \\ &\leq 2 \sum_{j=0}^{\infty} (j+1) \cdot q^{j(1-\frac{s}{2})} \end{aligned}$$

by Lemmas 4.16 and 4.17. This proves the absolute convergence of the series on the right hand side of (4.32) for $s \geq 3$. By a general theorem of harmonic analysis, this series therefore converges to $\theta(\tilde{t}; h, V)$ for all $\tilde{t} \in \mathcal{O}$.

For $s = 3$, we have

$$\theta(\tilde{t}; h, V) = \sum_{A/Q} \frac{\mu(Q)}{\Phi(Q)^3} \cdot E(-A\tilde{t}/Q) \tag{4.36}$$

as well as the infinite product representation (4.27). The summation in (4.36) is the global singular series for the polynomial 3-primes problem.

4.4.2 The derivative of $\gamma_1 X_1^{d_1} + \gamma_2 X_2^{d_2} + \cdots + \gamma_s X_s^{d_s}$ on $\mathcal{O}^s$

Let $\gamma_1, \gamma_2, \ldots, \gamma_s$ be non-zero constants in $\mathbb{F}_q$, and let $d_1, d_2, \ldots, d_s$ be strictly positive integers which are not divisible by p. As in §3.3, we define d by

$$\frac{s}{d} = \frac{1}{d_1} + \frac{1}{d_2} + \cdots + \frac{1}{d_s}.$$

By Theorems 3.12 and 3.13, the local derivatives $\theta_P(t_P; h_P, \mathcal{O}_P^s)$ for $P \in \mathbf{M}_{\text{fin}}$ of the diagonal polynomial

$$h(X_1, X_2, \ldots, X_s) = \gamma_1 X_1^{d_1} + \gamma_2 X_2^{d_2} + \cdots + \gamma_s X_s^{d_s} \qquad (4.37)$$

on $V_P = \mathcal{O}_P^s$ are continuous and satisfy the hypotheses of Theorem 4.12 whenever $s > 2d$. Therefore we have the following.

Theorem 4.18. *If $s > 2d$, the global Radon–Nikodym derivative of* (4.37) *on $V = \mathcal{O}^s$ exists as a continuous function on $\mathbb{O}$. Further,*

$$\theta(\tilde{t}; h, \mathcal{O}^s) = \prod_{P \in \mathbf{M}_{\text{fin}}} \theta_P(t_P; h_P, \mathcal{O}_P^s) \qquad (4.38)$$

where $\theta_P(t_P; h_P, \mathcal{O}_P^s)$ is defined by the local singular series (3.22).

We can also write down a global singular series for the derivative (4.38) using (4.23) and the local computations in §3.3. In order to show that this series converges to $\theta(\tilde{t}; h, \mathcal{O}^s)$ when $s > 2d$, we must first prove a lemma.

For any polynomial $Q \in \mathbf{A}$, let $\omega(Q)$ be the number of distinct, positive irreducibles which divide Q; and for a given degree r, let $\bar{\omega}(r)$ be the maximum value of $\omega(Q)$ on the polynomials of degree r.

Lemma 4.19. *For any $\epsilon > 0$, there is an $N(\epsilon)$ such that $\omega(Q) \leq \epsilon \cdot \deg(Q)$ for all Q with $\deg(Q) \geq N(\epsilon)$.*

Proof. It is clear that $\bar{\omega}(r)$ is an increasing function of r. Let $\delta(r)$ be the smallest δ such that

$$r \leq \gamma(\delta) = \sum_{i=1}^{\delta} i \cdot \pi(i),$$

where $\pi(i)$ is the number of positive irreducibles of degree i in $\mathbf{A}$. When $r = \gamma(\delta)$, the maximum value of $\omega(Q)$ clearly occurs at the polynomial Q_δ

which is the product of all the positive irreducibles of degree less than or equal to δ. Therefore for any r,

$$\bar{\omega}(r) \le \omega(Q_{\delta(r)}) = \sum_{i=1}^{\delta(r)} \pi(i) \le L_q(\delta(r)) \le \frac{3q}{2(q-1)} \cdot \frac{q^{\delta(r)}}{\delta(r)} \tag{4.39}$$

by (1.23). Now from the definition of $\delta(r)$,

$$r > \gamma(\delta(r)-1) = \sum_{i=1}^{\delta(r)-1} i \cdot \pi(i) \ge \sum_{i=1}^{\delta(r)-1} \frac{1}{2} \cdot q^i = \frac{q}{2} \cdot \frac{q^{\delta(r)-1}-1}{q-1} \tag{4.40}$$

by (1.23) again. It follows now from (4.39) and (4.40) that, for $\deg(Q) = r$,

$$\frac{\omega(Q)}{\deg(Q)} \le \frac{\bar{\omega}(r)}{r} \le \frac{3q}{\delta(r)} \cdot \left(1 - q^{1-\delta(r)}\right)^{-1}.$$

Since $\delta(r)$ goes to infinity with r, the right hand side of this inequality is less than any given $\epsilon > 0$ for sufficiently large r. ■

By (3.23) and the remarks preceding Theorem 3.13, the local Fourier transform $\hat{\theta}_P(A/Q; h_P, \mathcal{O}_P^s)$ satisfies the inequality

$$|\hat{\theta}_P(A/Q; h_P, \mathcal{O}_P^s)| \le (d_1 d_2 \cdots d_s) \cdot |P|^{-v_P(Q)s/d}$$

for every P dividing Q. Therefore, the global Fourier transform (4.23) satisfies

$$|\hat{\theta}(A/Q; h, \mathcal{O}^s)| \le (d_1 d_2 \cdots d_s)^{\omega(Q)} |Q|^{-s/d}.$$

Since there are $\Phi(Q) \le |Q|$ fractions A/Q with fixed denominator Q, this last inequality implies that

$$\begin{aligned} \sum_{A/Q} |\hat{\theta}(A/Q; h, \mathcal{O}^s)| &\le \sum_{r=0}^{\infty} \sideset{}{'}\sum_{\deg Q = r} |Q| (d_1 d_2 \cdots d_s)^{\omega(Q)} |Q|^{-s/d} \\ &\le \sum_{r=0}^{\infty} (d_1 d_2 \cdots d_s)^{\bar{\omega}(r)} \cdot q^{r(2-s/d)}. \end{aligned} \tag{4.41}$$

If $d_1 d_2 \cdots d_s = 1$, then the series (4.41) converges when $s > 2d$. Otherwise for $s > 2d$, let ϵ satisfy

$$0 < \epsilon < \frac{(s/d) - 2}{\log_q(d_1 d_2 \cdots d_s)}.$$

By Lemma 4.19, the summand in (4.41) is less than

$$q^{r(\epsilon \log_q(d_1 d_2 \cdots d_s) + 2 - s/d)}$$

for all $r \ge N(\epsilon)$, which shows that the global Fourier series (4.24) for $f(\tilde{t}) = \theta(\tilde{t}; h, \mathcal{O}^s)$ is absolutely convergent when $s > 2d$. By Theorem 4.18

and a general theorem of harmonic analysis, this Fourier series converges to $\theta(\tilde{t}; h, \mathcal{O}^s)$ for all $\tilde{t} \in \mathcal{O}$. We have therefore proved the following.

Theorem 4.20. *For $s > 2d$, the global Radon–Nikodym derivative* (4.38) *is represented by its global singular series* (4.24).

4.5 The idèle group of k

The *idèle group* J_k of k is the group of units of the adèle ring: $J_k = \mathbb{A}_k^\times$. We can also describe J_k as the restricted direct product over $\mathrm{v} \in \mathbf{M}$ of the multiplicative groups $k_\mathrm{v}^\times$ relative to their subgroups $\mathbf{U}_\mathrm{v}$. This means that J_k is the subgroup of the product $\prod_{\mathrm{v}\in\mathbf{M}} k_\mathrm{v}^\times$ which consists of the vectors $\mathbf{i} = (\mathbf{i}_\mathrm{v})_{\mathrm{v}\in\mathbf{M}}$ with $\mathbf{i}_\mathrm{v} \in \mathbf{U}_\mathrm{v}$ for almost all $\mathrm{v} \in \mathbf{M}$. We do not topologize J_k with the relative topology from $\mathbb{A}_k$, but rather we define the topology by giving a base $\mathcal{B}$ for the neighbourhoods of one:

$$B \in \mathcal{B} \iff B = \left(\prod_{\mathrm{v}\notin S} \mathbf{U}_\mathrm{v}\right) \times \left(\prod_{\mathrm{v}\in S} \mathbf{U}_\mathrm{v}^{(n_\mathrm{v})}\right) \tag{4.42}$$

where S is any finite subset of $\mathbf{M}$ and where each n_v, for $\mathrm{v} \in S$, is a positive integer. The group topology defined on J_k by $\mathcal{B}$ is locally compact because, e.g.,

$$W = \left(\prod_{\mathrm{v}\in\mathbf{M}_{\mathrm{fin}}} \mathbf{U}_\mathrm{v}\right) \times \mathbf{U}_\infty^{(1)} \in \mathcal{B}$$

is compact.

The multiplicative group $k_\mathrm{v}^\times$ of each completion k_v is embedded as a subgroup of J_k via the inclusion map $j_\mathrm{v} : k_\mathrm{v}^\times \longrightarrow J_k$, which is defined in the following way: for $t_\mathrm{v} \in k_\mathrm{v}^\times$, $j_\mathrm{v}(t_\mathrm{v})$ is the idèle which has v-component t_v and which has component one at every other place in $\mathbf{M}$. We can also embed $k^\times$ in J_k 'along the diagonal': the idèle associated to $x \in k^\times$ is the idèle which has component x at every place in $\mathbf{M}$. The topology induced on $k^\times$ as a subgroup of J_k is the discrete topology. To prove this fact, we show that $k \cap W = \{1\}$. Any element of $k^\times$ lying in W has no pole and must therefore be a constant. Since 1 is the only constant function lying in $\mathbf{U}_\infty^{(1)}$, $k \cap W = \{1\}$.

Just as we were able to decompose $\mathbb{A}_k$ as a direct product (Proposition 4.1), we can decompose J_k as a direct product of three of its subgroups.

Proposition 4.21. *We have $J_k = k^\times \cdot W \cdot T_\infty^{\mathbf{Z}}$, the internal direct product of topological groups, where $T_\infty = j_\infty(T)$ and where $T_\infty^{\mathbf{Z}}$ is the discrete cyclic group $\{T_\infty^n : n \in \mathbf{Z}\}$.*

Proof. Clearly each of the three groups $k^\times$ (embedded diagonally), W, and $T_\infty^{\mathbf{Z}}$ are subgroups of J_k, and it is easy to check that they intersect trivially in pairs. In fact, we have already shown that $k^\times \cap W = \{1\}$. Now let $\mathbf{i} \in J_k$ be arbitrary and let $S = \{\mathrm{v} \in \mathbf{M}_{\mathrm{fin}} : \mathbf{i}_\mathrm{v} \notin \mathbf{U}_\mathrm{v}\}$. Choosing positive irreducibles as uniformizers for all finite places, we set

$$x = \alpha \prod_{\mathrm{v} \in S} P_\mathrm{v}^{\mathrm{ord}_\mathrm{v}(\mathbf{i}_\mathrm{v})} \in k^\times ,$$

where $\alpha \in \mathbf{F}_q$ is the coefficient of the leading term of $\mathbf{i}_\infty$ in the Laurent expansion (2.3). The product makes sense because S is finite. Now for each $\mathrm{v} \in \mathbf{M}_{\mathrm{fin}}$, $\mathbf{i}_\mathrm{v}/x \in \mathbf{U}_\mathrm{v}$. Furthermore,

$$(\mathbf{i}/x)_\infty = u_\infty \cdot T_\infty^{\mathrm{ord}_\infty(\mathbf{i}_\infty/x)}$$

with $u_\infty \in \mathbf{U}_\infty^{(1)}$. Therefore we have

$$\mathbf{i} = x \cdot u \cdot T_\infty^{\mathrm{ord}_\infty(\mathbf{i}_\infty)}$$

with $u \in W$, as desired. ■

The group $\mathbf{C}_k = J_k/k^\times$ is called the *idèle class group*. One immediate implication of Proposition 4.21 is that, unlike its sister the adèle class group, $\mathbf{C}_k$ is not compact. We proceed now to introduce a compact quotient of J_k which will be of fundamental importance for our work in Chapter 5.

The *divisor group* of k is the free abelian group $\mathfrak{D}_k$ on the set of places $\mathbf{M}$ of k. Since we wish to write $\mathfrak{D}_k$ in multiplicative notation, a place $\mathrm{v} \in \mathbf{M}$ viewed as an element of $\mathfrak{D}_k$ will be called a *prime divisor* and will be denoted by '$\mathfrak{p}_\mathrm{v}$'. A typical divisor $\mathfrak{d} \in \mathfrak{D}_k$ then has the form

$$\mathfrak{d} = \prod_{\mathrm{v} \in \mathbf{M}} \mathfrak{p}_\mathrm{v}^{e_\mathrm{v}} \tag{4.43}$$

where $e_\mathrm{v} \in \mathbf{Z}$ for all $\mathrm{v} \in \mathbf{M}$ and where $e_\mathrm{v} = 0$ for almost all $\mathrm{v} \in \mathbf{M}$. For $\mathrm{v} \in \mathbf{M}$, we define $\deg(\mathrm{v}) = \deg(\mathfrak{p}_\mathrm{v})$ to be the $\mathbf{F}_q$-dimension of the residue class field $\mathbf{K}(\mathrm{v})$. The integer $\deg(\mathrm{v})$ is often called the 'residual degree'. It equals one when $\mathrm{v} = \infty$ and equals $\deg(P_\mathrm{v})$ for $\mathrm{v} \in \mathbf{M}_{\mathrm{fin}}$, where P_v is the positive irreducible polynomial which serves as our standard uniformizer for k_v. We put

$$|\mathfrak{p}_\mathrm{v}| = q^{\deg(\mathfrak{p}_\mathrm{v})} .$$

These definitions extend by linearity from prime divisors to any divisor (4.43):

$$\deg(\mathfrak{d}) = \sum_{\mathrm{v} \in \mathbf{M}} e_\mathrm{v} \deg(\mathfrak{d}_\mathrm{v})$$

and

$$|\mathfrak{d}| = \prod_{v \in M} |\mathfrak{p}_v|^{e_v} = q^{\deg(\mathfrak{d})} . \tag{4.44}$$

Now every idèle $\mathbf{i} = (\mathbf{i}_v)_{v \in M}$ has a divisor $\partial(\mathbf{i})$ associated to it in a natural way:

$$\partial(\mathbf{i}) = \prod_{v \in M} \mathfrak{p}_v^{v(\mathbf{i}_v)}$$

the product being actually finite since $v(\mathbf{i}_v) = 0$ for almost $v \in M$. The degree of $\mathbf{i}$ is then defined to be the degree of its divisor $\partial(\mathbf{i})$:

$$\deg(\mathbf{i}) = \deg(\partial(\mathbf{i})) = \sum_{v \in M} \deg(v) \cdot v(\mathbf{i}_v) .$$

Since $k^\times \subset J_k$, $\deg(x)$ is defined for every $x \in k^\times$. In fact, as we see immediately from (2.1), $\deg(x) = 0$ for every $x \in k^\times$. This fact justifies the following definition.

Definition 4.22. *Let* $J_k^0 = \{\mathbf{i} \in J_k : \deg(\mathbf{i}_v) = 0\}$ *be the kernel of the degree map on idèles. The group* $\mathbf{C}_k^0 = J_k^0/k^\times$ *is called the* idèle class group of degree *zero.*

Proposition 4.23. *The group* $\mathbf{C}_k^0$ *is algebraically and topologically isomorphic to* W. *In particular,* $\mathbf{C}_k^0$ *is compact.*

Proof. The degree function on idèles vanishes on $k^\times$ and W whereas $\deg(T_\infty) = 1$. Therefore Proposition 4.21 implies that $J_k^0 = k^\times W$. Since W is open in J_k^0, the results follow. ■

5
L-functions of Dirichlet type

In classical number theory, the analysis of problems involving prime numbers often leads to questions about the Riemann zeta-function or its generalizations, the Dirichlet L-functions. In this chapter, we develop a particular class of L-functions over k and use them to prove a bound for character sums (Theorem 5.7) which will be of fundamental importance in our analysis of the polynomial 3-primes problem.

5.1 Multiplicative characters and L-functions

We begin by studying the continuous characters of the idèle group J_k. For $\mathrm{v} \in \mathbf{M}$, let $j_\mathrm{v} : k_\mathrm{v}^\times \longrightarrow J_k$ be the inclusion map defined in §4.5. Then for a given character χ of J_k,

$$\chi_\mathrm{v} = \chi \circ j_\mathrm{v} \in \hat{k}_\mathrm{v}^\times$$

is called the *local component* of χ at v.

Proposition 5.1. *Let $\chi \in \hat{J}_k$. Then for almost all $\mathrm{v} \in \mathbf{M}$, the conductor $m(\chi_\mathrm{v}) = 0$.*

Proof. The base $\mathcal{B}$ of neighbourhoods of unity in J_k defined by (4.42) consists of open subgroups. Since there are no non-trivial subgroups of the circle group $\mathbb{T}$ in the open right half-plane, $\ker(\chi)$ must contain one of the subgroups $\mathrm{B} \in \mathcal{B}$. Since B contains $j_\mathrm{v}(\mathbf{U}_\mathrm{v})$ for almost all $\mathrm{v} \in \mathbf{M}$, the result follows. ∎

This proposition shows that χ can be decomposed into local factors. In fact since χ is continuous,

$$\chi(t) = \prod_{\mathrm{v} \in \mathbf{M}} \chi_\mathrm{v}(t_\mathrm{v}) \tag{5.1}$$

for all $t = (t_\mathrm{v})_{\mathrm{v}\in\mathbf{M}} \in J_k$, the product on the right being actually finite. Conversely if $\chi_\mathrm{v} \in \hat{k}_\mathrm{v}^\times$ is a sequence of local characters such that $m(\chi_\mathrm{v}) = 0$ for almost all $\mathrm{v} \in \mathbf{M}$, then (5.1) defines a character χ of J_k.

For a 'global' character, it is convenient to define the conductor at a place $\mathrm{v} \in \mathbf{M}$ to be a divisor rather than an integer. For $\chi \in \hat{J}_k$, we write

$$\mathfrak{m}_{\mathrm{v}}(\chi) = \mathfrak{p}_{\mathrm{v}}^{m(\chi_{\mathrm{v}})}$$

and we call $\mathfrak{m}_{\mathrm{v}}(\chi)$ the *local conductor* of χ at v. We say that χ is *unramified* at v if $\mathfrak{m}_{\mathrm{v}}(\chi) = 1$; otherwise χ is *ramified* at v. If $\mathfrak{m}_{\mathrm{v}}(\chi) = \mathfrak{p}_{\mathrm{v}}$, then χ is said to be *tamely ramified* at v; otherwise, χ is *wildly ramified* at v.

By Proposition 5.1, χ is unramified at almost all $\mathrm{v} \in \mathbf{M}$. This fact justifies the following definition.

Definition 5.2. *For $\chi \in \hat{J}_k$, the* global conductor *of χ is the divisor*

$$\mathfrak{m}(\chi) \stackrel{\text{def}}{=} \prod_{\mathrm{v} \in \mathbf{M}} \mathfrak{m}_{\mathrm{v}}(\chi) .$$

Let S be any finite set of k-places containing all those places where χ ramifies. For $\mathrm{v} \notin S$, choose a uniformizer $\pi_{\mathrm{v}} \in k_{\mathrm{v}}$ and define

$$\chi_*(\mathfrak{p}_{\mathrm{v}}) \stackrel{\text{def}}{=} \chi_{\mathrm{v}}(\pi_{\mathrm{v}}) . \tag{5.2}$$

This definition is independent of the choice of uniformizer since χ is unramified at v. Let $\mathfrak{D}_S \subseteq \mathfrak{D}_k$ (see §4.5) be the subgroup consisting of the divisors which are prime to S. Since $\mathfrak{D}_S$ is a free abelian group, Definition 5.2 extends to a homomorphism $\chi_* : \mathfrak{D}_S \longrightarrow \mathbf{T}$.

Now let

$$S(\chi) = \mathrm{Supp}(\mathfrak{m}(\chi)) = \text{ the set of places dividing } \mathfrak{m}(\chi)$$

and let $\mathcal{M}(\chi)$ be the monoid of *integral* divisors in $\mathfrak{D}_{S(\chi)}$. For $s \in \mathbf{C}$, $\mathrm{Re}(s) > 1$, we define

$$L(s,\chi) \stackrel{\text{def}}{=} \prod_{\mathrm{v} \notin S(\chi)} \left(1 - \chi_*(\mathfrak{p}_{\mathrm{v}})|\mathfrak{p}_{\mathrm{v}}|^{-s}\right)^{-1} \tag{5.3}$$

$$= \sum_{\mathfrak{d} \in \mathcal{M}(\chi)} \frac{\chi_*(\mathfrak{d})}{|\mathfrak{d}|^s} \tag{5.4}$$

(see §4.5). The function $L(s,\chi)$ is called the *L-function* of χ, and it is represented both by its *Euler product* (5.3) and its *Dirichlet series* (5.4). The product and the series converge uniformly on compact subsets of the open half-plane $\mathrm{Re}(s) > 1$ and therefore define $L(s,\chi)$ as a holomorphic function in that half-plane. When χ is the trivial character χ_0 on J_k, $S(\chi_0)$

is empty, and $L(s, \chi_0)$ is the *zeta-function* $Z_k(s)$ of the rational function field k. One can sum the Dirichlet series (5.4) for $\chi = \chi_0$ in closed form:

$$Z_k(s) = \prod_{\mathrm{v} \in \mathbf{M}} \left(1 - |\mathfrak{p}_{\mathrm{v}}|^{-s}\right)^{-1} = \frac{1}{(1 - q^{-s})(1 - q^{1-s})} \cdot \qquad (5.5)$$

We now restrict our attention to a particular class of characters on J_k. We say that a character $\chi \in \hat{J}_k$ is *arithmetic* if χ arises by pull-back from a character of the idèle class group $\mathbf{C}_k$. Thus, χ is arithmetic if and only if $\chi(k^\times) = 1$. Class field theory tells us that such characters arise in a natural way from cyclic extension fields of k.

Definition 5.3. *An arithmetic character χ is* of Dirichlet type *(relative to the choice of generator T of k) if $\chi_\infty(T_\infty) = 1$, where $T_\infty = j_\infty(T)$. An L-function $L(s, \chi)$ is said to be* of Dirichlet type *if the character χ used to define it is of Dirichlet type.*

It is clear from Propositions 4.21 and 4.23 that the group of characters of Dirichlet type is naturally identified with $\hat{J}_k^0 \cong \widehat{W}$.

The reader may find the terminology 'of Dirichlet type' puzzling in this context and may ask 'How are these characters related to the usual Dirichlet characters which are familiar from the classical theory over $\mathbb{Q}$?' The remainder of this section is devoted to answering this question for characters of Dirichlet type which are either unramified or tamely ramified at ∞. These are the characters which are the strict analogues over k of the Dirichlet characters over $\mathbb{Q}$.

Every integral divisor $\mathfrak{d} \in \mathfrak{D}$ which is prime to ∞ is the divisor of zeros of a unique positive polynomial $D \in \mathbf{A}$, and we will identify $\mathfrak{d}$ and D. In particular, if χ is a given character of Dirichlet type, then

$$\mathfrak{m}(\chi) = \infty^{\ell+1} Q \qquad (5.6)$$

for some $\ell \geq -1$ and some positive polynomial Q. Put $S = S(\chi) \cup \{\infty\}$. Then from the definitions, for a positive polynomial D which is prime to Q,

$$1 = \chi(D) = \chi_*(D) \prod_{\mathrm{v} \in S} \chi_{\mathrm{v}}(D_{\mathrm{v}})$$

(the first equality since χ is arithmetic); and therefore

$$\chi_*(D) = \prod_{\mathrm{v} \in S} \chi_{\mathrm{v}}^{-1}(D_{\mathrm{v}}) \, . \qquad (5.7)$$

Let us assume that χ is at most tamely ramified at ∞, i.e. that $\ell = -1$ or 0. Therefore, since $D_\infty / T_\infty^{\deg D} \in \mathbf{U}_\infty^{(1)}$, we have

$$\chi_\infty(D_\infty) = \chi_\infty(D_\infty/T_\infty^{\deg D}) = 1 \,. \tag{5.8}$$

We see now from (5.7) that $\chi_*(D) = 1$ whenever $D \equiv 1 \pmod{Q}$, because then $v(D_v) \geq m(\chi_v)$ for every $v \in S(\chi)$. This means then that we can view χ_* as a character of the group of units of the quotient ring $\mathbf{A}/Q\mathbf{A}$, because the positive polynomials of degree $\deg(Q)$ form a complete residue system modulo Q. In particular, $\chi_*(\alpha)$ is defined for all $\alpha \in \mathbf{F}_q^\times$.

Proposition 5.4. *Suppose that χ is at most tamely ramified at ∞. Then χ is unramified at ∞ (i.e. $\ell = -1$) if and only if $\chi_*(\mathbf{F}_q^\times) = 1$.*

Proof. Given $\alpha \in \mathbf{F}_q^\times$, let D be a positive polynomial such that $D \equiv \alpha \pmod{Q}$. By (5.7) and (5.8),

$$\chi_*(\alpha) = \chi_*(D) = \prod_{v \in S - \{\infty\}} \chi_v^{-1}(D_v) = \prod_{v \in S - \{\infty\}} \chi_v^{-1}(\alpha) \,.$$

On the other hand,

$$1 = \chi(\alpha) = \prod_{v \in S} \chi_v(\alpha)$$

since χ is unramified outside of S. Hence

$$\chi_\infty(\alpha) = \prod_{v \in S - \{\infty\}} \chi_v^{-1}(\alpha) = \chi_*(\alpha) \,,$$

and so $\chi_*(\mathbf{F}_q^\times) = 1$ if and only if χ_∞ is trivial on $\mathbf{U}_\infty$. ∎

A Dirichlet character χ_* over $\mathbf{Q}$ is unramified at the archimedean place of $\mathbf{Q}$ (i.e. the 'place at infinity') if and only if it is an *even* character, which means that $\chi_*(-1) = 1$. Since $\mathbf{Z}^\times = \{\pm 1\}$, a Dirichlet character is even when it is trivial on the units of $\mathbf{Z}$. Because $\mathbf{F}_q^\times$ is the group of units of the polynomial ring $\mathbf{A}$, Proposition 5.4 makes an analogous statement for the characters of Dirichlet type over k which are unramified at ∞.

Since the infinite place of $\mathbf{Q}$ is archimedean, the behaviour of a Dirichlet character at infinity is rather limited. It can ramify or not ramify, but it cannot ramify wildly. The characters of Dirichlet type over k have a much more interesting behaviour at ∞. The meaning of this behaviour for the arithmetic of $\mathbf{A}$ is explored in Exercises 5.1(2) and 5.2(3).

Exercises 5.1

1. Prove (5.5). [Hint: Use the Dirichlet series for $L(s, \chi_0)$. Divisors which are prime to ∞ correspond to positive polynomials, which can be grouped by degree.] Use your results to prove that the Dirichlet

series (5.4) for $L(s,\chi)$ converges uniformly in the half-plane $\mathrm{Re}(s) \geq 1+\varepsilon$ for every $\varepsilon > 0$. Prove further that the Euler product (5.3) and the Dirichlet series (5.4) represent the same function in the half-plane $\mathrm{Re}(s) > 1$.

2. Let $\mathcal{P}$ be the monoid of all positive polynomials, and let $A, B \in \mathcal{P}$ have degrees a and b respectively. For $\ell \geq 0$, we say that A and B *have the same first ℓ coefficients* if the coefficients of the monomials T^{a-i} and T^{b-i} in A and B respectively are equal for $i = 0, 1, \ldots, \ell$. Given $\ell \geq 0$ and $Q \in \mathcal{P}$, define a relation $\mathbf{R}_{\ell,Q}$ on $\mathcal{P}$ as follows: for $A, B \in \mathcal{P}$
$$A \equiv B \pmod{\mathbf{R}_{\ell,Q}} \tag{5.9}$$
if and only if $A \equiv B \pmod{Q}$ and further A and B have the same first ℓ coefficients. Show that $\mathbf{R}_{\ell,Q}$ is a *congruence relation* on $\mathcal{P}$ (i.e. $\mathbf{R}_{\ell,Q}$ is an equivalent relation satisfying
$$A \equiv B \pmod{\mathbf{R}_{\ell,Q}} \Rightarrow AC \equiv BC \pmod{\mathbf{R}_{\ell,Q}}$$
for all $A, B, C \in \mathcal{P}$). Show also that if χ is a character of Dirichlet type with conductor $\mathfrak{m}(\chi)$ dividing $\infty^{\ell+1}Q$ and if $\chi_*(D)$ is defined as above, then (5.9) implies $\chi_*(A) = \chi_*(B)$ whenever A and B are prime to Q. Therefore, χ_* can be regarded as a character of the group of units $G_{\ell,Q}$ of the quotient monoid $\mathcal{P}/\mathbf{R}_{\ell,Q}$. Show that $G_{\ell,Q}$ is finite of order $q^\ell\Phi(Q)$, where $\Phi(Q)$ is the order of the group of units of $\mathbf{A}/Q\mathbf{A}$. Show further that every character of $G_{\ell,Q}$ is induced in this way by a character of Dirichlet type.

5.2 The Riemann hypothesis

Throughout this section, χ will be a character of Dirichlet type on J_k. Our aim is to derive a sharp bound for the absolute value of the character sum
$$\pi_\nu(\chi) \stackrel{\text{def}}{=} \sum_{\deg P=\nu}{}' \chi_*(P) \tag{5.10}$$
where $\nu > 0$ and where the summation is over the positive irreducibles P of degree ν and prime to the finite part of $\mathfrak{m}(\chi)$. When $\chi = \chi_0$, this sum equals the number of positive irreducibles of degree ν. As is well known,
$$\sum_{\mu\tau=\nu} \mu \cdot \pi_\mu(\chi_0) = q^\nu \Rightarrow \pi_\nu(\chi_0) \leq \frac{q^\nu}{\nu}\,, \tag{5.11}$$
the first equality being essentially equivalent to (5.5). In order to bound $\pi_\nu(\chi)$ when $\chi \neq \chi_0$, we will need to investigate the properties of $L(s,\chi)$ as an analytic function of s. We put $d(\chi) = \deg(\mathfrak{m}(\chi)) - 2$.

Theorem 5.5. *If χ is a non-trivial character of Dirichlet type, then*

$$L(s,\chi) = P_\chi(q^{-s})$$

where $P_\chi(z) \in \mathbb{C}[z]$ has degree $d(\chi)$ and $P_\chi(0) = 1$. In particular, $L(s,\chi)$ is an entire *complex analytic function of s.*

This result tells us that $L(s,\chi)$ is in fact a polynomial in q^{-s}. The reader should contrast this result with (5.5). Although it is possible to prove Theorem 5.5 in an elementary way, we will derive it and the next theorem from results of A. Weil on the zeta-functions and L-functions associated to algebraic curves over $\mathbb{F}_q$.

Theorem 5.6. *If χ is a non-trivial character of Dirichlet type, then the L-function $L(s,\chi)$ satisfies the following analogue of the Riemann hypothesis: suppose*

$$P_\chi(z) = \prod_{i=1}^{d(\chi)} (1 - \gamma_i z)$$

in $\mathbb{C}[z]$. Then for $1 \leq i \leq d(\chi)$,

$$|\gamma_i| = q^{1/2}\ . \tag{5.12}$$

Proofs. By class field theory, $L(s,\chi)$ is the Artin L-series of a character of the Galois group of a cyclic extension K/k; and $L(s,\chi)$ divides the zeta-function $Z_K(s)$ of K. Since $\chi_\infty(T_\infty) = 1$, this extension is geometric (i.e. it contains no elements algebraic over $\mathbb{F}_q$ other than the elements of $\mathbb{F}_q$ itself). We may now appeal to the famous results of Weil (1948) to conclude that Theorems 5.5 and 5.6 are indeed valid. ■

To the reader who is unfamiliar with the existence theorem of class field theory or the above results of Weil, we apologize for including a proof which requires a degree of background well beyond what is assumed at any other point in this monograph. We sketch a proof of Theorem 5.5 in Exercise 3 below. However, Theorem 5.6 is a deep result and so requires some effort for its proof. There now exists an elegant and relatively accessible proof of the Riemann hypothesis for the zeta-function $Z_K(s)$ of a finite extension field K of k, due to Bombieri (1974). However, the reader who works through that proof will still need to know how to construct a geometric extension K/k having the property that $L(s,\chi)$ divides $Z_K(s)$. In fact, this can be done explicitly without class field theory using a construction of Carlitz (cf. Hayes 1974). We hope these remarks will be helpful to the reader of limited background who wishes to gain insight into why Theorems 5.5 and 5.6 are true.

We are now in a position to prove our main result.

Theorem 5.7. *If χ is a non-trivial character of Dirichlet type, then*

$$|\pi_\nu(\chi)| \leq \{\deg(\mathfrak{m}(\chi)) - \lambda(\chi) + C_q\} \cdot \frac{q^{\nu/2}}{\nu} \tag{5.13}$$

where $\lambda(\chi) = 2$ if χ is ramified at ∞ and $\lambda(\chi) = 1$ if χ is unramified at ∞. The constant C_q is independent of both ν and χ. Further, as q gets large, C_q decreases down to one.

Proof. For $S = S(\chi) \cup \{\infty\}$, put

$$\begin{aligned} L_S(s,\chi) &\stackrel{\text{def}}{=} \prod_{\mathrm{v}\notin S} (1 - \chi_*(\mathfrak{p}_\mathrm{v})|\mathfrak{p}_\mathrm{v}|^{-s})^{-1} \\ &= \prod_{P} \left(1 - \chi_*(P) \cdot q^{-s\cdot\deg(P)}\right)^{-1} \end{aligned} \tag{5.14}$$

where P runs through all the positive irreducible polynomials which are prime to the finite part of $\mathfrak{m}(\chi)$. An elementary calculation using the logarithmic derivative gives us

$$\frac{1}{\log(q)} \cdot \frac{L_S'(s,\chi)}{L_S(s,\chi)} = -\sum_{\nu=1}^{\infty} \left(\sum_{\mu\tau=\nu} \mu \sum_{\deg P=\mu}' \chi_*(P^\tau) \right) q^{-s\nu}. \tag{5.15}$$

On the other hand if $a(\chi) = \deg(\mathfrak{m}(\chi)) - \lambda(\chi)$, then

$$L_S(s,\chi) = \prod_{i=1}^{a(\chi)} (1 - \beta_i q^{-s})$$

with $|\beta_i| \leq q^{1/2}$ for $1 \leq i \leq a(\chi)$. When $\lambda(\chi) = 2$, this is precisely the claim of Theorem 5.6 (since $L_S = L$ in this case); but it is also true when $\lambda(\chi) = 1$ because then

$$L_S(s,\chi) = (1 - \chi_*(\infty)|\infty|^{-s}) \cdot L(s,\chi) = (1 - q^{-s})P_\chi(q^{-s})$$

since χ is unramified at ∞ and $\chi_*(\infty) = \chi_\infty(1/T_\infty) = 1$. Thus we have also

$$\begin{aligned} \frac{1}{\log(q)} \cdot \frac{L_S'(s,\chi)}{L_S(s,\chi)} &= \sum_{i=1}^{a(\chi)} \frac{\beta_i q^{-s}}{1 - \beta_i q^{-s}} \\ &= \sum_{i=1}^{a(\chi)} \sum_{\nu=1}^{\infty} \beta_i^\nu q^{-\nu s} \end{aligned}$$

$$= \sum_{\nu=1}^{\infty} \left(\sum_{i=1}^{a(\chi)} \beta_i^{\nu} \right) q^{-\nu s} .$$

Comparing coefficients on the right hand sides of this equation and (5.15), we see that

$$\sum_{\mu\tau=\nu} \mu \sum_{\deg P=\mu}{}' \chi_*(P^{\tau}) = -\sum_{i=1}^{a(\chi)} \beta_i^{\nu} . \tag{5.16}$$

Now each β_i is bounded in absolute value by $q^{1/2}$; and by (5.11) each summation

$$\sum_{\deg P=\mu}{}' \chi_*(P^{\tau})$$

is trivially bounded in absolute value by $\pi_\mu(\chi_0) \leq q^{\mu}/\mu$. Therefore (5.16) implies the bound

$$| \nu \cdot \pi_\nu(\chi) | \leq a(\chi) \cdot q^{\nu/2} + \sum_{\substack{\mu\tau=\nu \\ \tau\neq 1}} q^{\mu} .$$

Since

$$\sum_{\substack{\mu\tau=\nu \\ \tau\neq 1}} q^{\mu} \leq q^{\nu/2} + q^{\nu/3} \sum_{\mu\leq\nu/3} 1 \leq q^{\nu/2} + \left(\frac{\nu}{3}\right) q^{\nu/3} ,$$

we conclude that

$$| \pi_\nu(\chi) | \leq (a(\chi) + C_q)\frac{q^{\nu/2}}{\nu}$$

where C_q is the maximum value of the function $1 + (x/3) \cdot q^{-x/6}$ on the positive real numbers. A standard calculation shows that this maximum occurs at $x = 6/\log(q)$. Therefore,

$$C_q = 1 + \frac{2}{e \cdot \log(q)} ,$$

a bounded function of q which approaches 1 as q becomes infinitely large. ■

Exercises 5.2

1. By (5.5),

$$\prod_P \left(1 - |P|^{-s}\right)^{-1} = \frac{1}{1 - q^{1-s}}$$

where P runs through the set of positive irreducibles in **A**. Use this identity to prove the equality in (5.11).

2. Prove Theorem 5.5 as follows: for $S = S(\chi) \cup \{\infty\}$, show from the Dirichlet series of (5.14) that

$$L_S(s,\chi) = \sum_{\nu=0}^{\infty} \sigma_\nu(\chi) q^{-\nu s}$$

where

$$\sigma_\nu(\chi) = \sum_{\deg D = \nu}{}' \chi_*(D) ,$$

the sum being taken over all positive polynomials D of degree ν which are prime to the polynomial Q of (5.6). Let $j = \deg(Q)$. Show that when $\nu > \ell + j - 1$, the positive polynomials of degree ν constitute a complete set of representatives for the congruence classes modulo $\mathbf{R}_{\ell,Q}$ of Exercise 5.1(2). Conclude that $\sigma_\nu(\chi) = 0$ for $\nu > \ell + j - 1$ so that $L_S(s,\chi) = P_S(q^{-s})$ where

$$P_S(u) = \sum_{\nu=0}^{\ell+j-1} \sigma_\nu(\chi) u^\nu .$$

When $\ell > 0$, $P_\chi(u) = P_S(u)$. When $\ell = 0$, $P_\chi(u) = P_S(u)/(1-u)$. Use Proposition 5.4 to show that $P_S(1) = 0$ when $\ell = 0$.

3. Let $\ell \geq 0$ and $Q \in \mathbf{A}$ be fixed, and let A be a positive polynomial which is prime to Q. Let $\pi_\nu(A;\ell,Q)$ be the number of positive irreducibles P of degree ν such that $P \equiv A \pmod{\mathbf{R}_{\ell,Q}}$ (see Exercise 5.1(2)). Using (5.13), show that

$$\pi_\nu(A;\ell,Q) = \frac{q^{\nu-\ell}}{\nu\Phi(Q)} + O\left(\frac{q^{\theta\nu}}{\nu}\right) \tag{5.17}$$

as ν approaches infinity, where $\theta = 1/2$ and where $\Phi(Q)$ is the order of the group of units of $\mathbf{A}/Q\mathbf{A}$.

5.3 A brief historical note

The L-functions of Dirichlet type were first introduced by Kornblum (1919) in order to prove the polynomial analogue of Dirichlet's theorem on primes in arithmetic progressions.

Theorem 5.8. (Kornblum) *Let Q be a positive polynomial in $\mathbf{A}$, and let A be any polynomial which is prime to Q. Then there are infinitely many positive irreducibles P such that $P \equiv A \pmod{Q}$.*

Actually, Kornblum introduced only the L-functions associated to characters which are at most tamely ramified at ∞ (i.e. $\ell = -1$ or 0 in (5.6)) because only these L-functions were needed for his theorem. After completing a first draft of his dissertation, Kornblum volunteered for service in World War I and was killed in action in 1914 at the age of 24. His paper was edited and published by his dissertation director, E. Landau.

Artin refined Kornblum's work in his own dissertation published in 1924. He proved the asymptotic formula (5.17) in the case $\ell = 0$ (and $q = p$) for some undetermined value θ satisfying $\frac{1}{2} \leq \theta < 1$ and independent of ν. Artin's main goal was to develop the theory of 'quadratic function fields' (i.e. extensions K/k of degree 2) and their zeta-functions $Z_K(s)$. He proved this refinement of Kornblum's theorem as an application of his results on zeta-functions of 'real quadratic' function fields.

In this same paper, Artin introduced the idea that the Riemann hypothesis is valid for the zeta-functions $Z_K(s)$. If true, this would imply the Riemann hypothesis for the L-functions of Dirichlet type associated to quadratic characters χ. Artin computed $P_\chi(z)$ in many examples with $p = 3, 5$, or 7 for such quadratic characters. He verified that the Riemann hypothesis was indeed true in all his examples. However, the Riemann hypothesis for function fields was not proved in general until 1948 when Weil's great memoir appeared. The reader interested in learning more about the early history of the Riemann hypothesis can profitably consult §4.10 and §5.8 in Kapital V of Eichler (1963).

Special cases of (5.17) with $\ell \geq 1$ were proved by Carlitz and Uchiyama in 1952 and 1955 respectively. Hayes proved (5.17) for arbitrary ℓ and Q in 1965. Uchiyama and Hayes do not invoke the Riemann hypothesis and therefore only assert that the asymptotic formula holds for some fixed value θ satisfying $\frac{1}{2} \leq \theta < 1$ and independent of ν. The validity of (5.17) with $\theta = \frac{1}{2}$ was first demonstrated by Rhin in 1972.

Exercises 5.3

1. Prove the second theorem of Kornblum (1919): with hypotheses as in Theorem 5.8, let d and n be positive integers with $n > 1$. Then there are infinitely many positive irreducibles P such that $P \equiv A \pmod{Q}$ and $\deg(P) \equiv d \pmod{n}$.

5.4 A lower bound for the 3-primes singular series

From the product representation (4.27), the singular series $\theta(\tilde{t}; h, V)$ of §4.4.1 vanishes at $\tilde{t} \in \mathcal{O}$ if and only if there is a $P \in \mathbf{M}_{\text{fin}}$ with $|P| = 2$ such that $\mathrm{v}_P(\tilde{t}_P) > 0$. If $q = 2$, let

$$\mathcal{E}_q \stackrel{\text{def}}{=} \mathbf{U}_T \times \mathbf{U}_{T+1} \times \prod_{\substack{P \in \mathbf{M}_{\text{fin}} \\ P \neq T \\ P \neq T+1}} \mathcal{O}_P$$

and otherwise let $\mathcal{E}_q \stackrel{\text{def}}{=} \mathcal{O}$. The singular series is a strictly positive function on $\mathcal{E}_q$. Therefore since $\mathcal{E}_q$ is compact, there is a constant κ_q such that

$$\theta(\tilde{t}; h, V) \geq \kappa_q > 0$$

for all $\tilde{t} \in \mathcal{E}_q$. In order to compute an explicit error term for the polynomial 3-primes problem, we need explicit numerical values for these κ_q. The closed form (5.5) for the zeta-function of k provides us with a tool for computing such values.

Proposition 5.9. *If* $q \geq 3$, *then*

$$\kappa_q = (1 - 1/q) \cdot \left(\frac{1 - 1/q^2}{1 - 1/q^5}\right)^5 \tag{5.18}$$

is a lower bound for $\theta(\tilde{t}; h, V)$.

Proof. For $\tilde{t} \in \mathcal{E}_q = \mathcal{O}$, the product (4.27) implies that

$$\theta(\tilde{t}; h, V) \geq \prod_{P \in \mathbf{M}_{\text{fin}}} \left(1 - \frac{1}{(|P| - 1)^2}\right). \tag{5.19}$$

We will employ the inequality

$$1 - \frac{1}{(x-1)^2} \geq \frac{(1 - x^{-6})^{-5}}{(1 - x^{-2})^{-1} \cdot (1 - x^{-3})^{-5}} \tag{5.20}$$

which is valid for all $x \geq 3$. This inequality may be derived as follows: for $x \geq 3$,

$$\begin{aligned} \frac{1 - \frac{1}{x^2}}{1 - \frac{1}{(x-1)^2}} &= 1 + \frac{2}{x^3}\left(1 + \frac{3/2}{x-2}\right) \leq 1 + \frac{5}{x^3} \\ &\leq \left(1 + \frac{1}{x^3}\right)^5 = \left(\frac{1 - x^{-6}}{1 - x^{-3}}\right)^5. \end{aligned}$$

Substituting $x = |P|$ in (5.20) and taking the product of both sides over all $P \in \mathbf{M}_{\text{fin}}$, we conclude that

$$\prod_{P\in\mathbf{M}_{\mathrm{fin}}}\left(1-\frac{1}{(|P|-1)^2}\right) \geq \frac{Z_k^*(6)^5}{Z_k^*(2)\cdot Z_k^*(3)^5} \tag{5.21}$$

where

$$Z_k^*(s) = \left(1-q^{-s}\right)\cdot Z_k(s) = \frac{1}{1-q^{1-s}} \tag{5.22}$$

is the zeta-function of k with its Euler factor at $\mathrm{v}=\infty$ removed. The second equality in (5.22) follows from (5.5). We see from this last equation and (5.19) that (5.21) implies (5.18). ■

By the usual methods, one can show that the function of q on the right hand side of (5.18) is increasing for $q \geq 3$. Since its value at $q = 3$ is about 0.3777, we conclude that

$$\theta(\tilde{t}; h, V) \geq 0.37 \tag{5.23}$$

for all $q \geq 3$. The 'usual methods', in their modern form, involve simply drawing a graph of the function over the interval $[3, 29]$ using a computer graphics package. We did so and observed that the function does indeed increase. The interested reader can obtain the same result using calculus.

Proposition 5.10. *For $q = 2$, $\kappa_2 = 3.21$ is a lower bound for $\theta(\tilde{t}; h, V)$ on $\mathcal{E}_2$.*

Proof. Let $S = \mathbf{M}_{\mathrm{fin}} - \{T, T+1\}$. For $\tilde{t} \in \mathcal{E}_2$, we have

$$\begin{aligned}\theta(\tilde{t}; h, V) &\geq 4\prod_{P\in S}\left(1-\frac{1}{(|P|-1)^2}\right)\\ &\geq 4\prod_{P\in S}\frac{(1-|P|^{-6})^{-5}}{(1-|P|^{-2})^{-1}\cdot(1-|P|^{-3})^{-5}}\end{aligned}$$

by (5.20) since $|P| > 3$ for $P \in S$. After we adjust for the eight missing Euler factors, the right hand side above equals

$$\frac{4\cdot(1-2^{-6})^{10}\cdot Z_k^*(6)^5}{(1-2^{-2})^2\, Z_k^*(2)\cdot(1-2^{-3})^{10}\, Z_k^*(3)^5} = 3.2113\,.$$

■

Putting this last proposition together with (5.23), we have finally

Proposition 5.11. *For any prime power q and any $\tilde{t} \in \mathcal{E}_q$,*

$$\theta(\tilde{t}; h, V) \geq 0.37\,. \tag{5.24}$$

Exercises 5.4

1. If f is a function from the positive polynomials of **A** into $\mathbb{C}$, we may define the formal Dirichlet series

$$Z_f(s) \stackrel{\text{def}}{=} \sum_Q{}' \frac{f(Q)}{|Q|^s} = \sum_{j=0}^{\infty} q^{-js} \sum_{\deg Q=j}{}' f(Q), \qquad (5.25)$$

Q running over all positive polynomials in **A**. If f is multiplicative (i.e. Q_1, Q_2 relatively prime $\Rightarrow f(Q_1Q_2) = f(Q_1)f(Q_2)$), then as formal Dirichlet series

$$Z_f(s) = \prod_P \sum_{r=0}^{\infty} \frac{f(P^r)}{|P|^{rs}} \qquad (5.26)$$

where the product is over all positive irreducibles in **A**. Use this identity together with (5.22) to show that

$$Z_\tau(s) = \left(1 - q^{1-s}\right)^{-2}$$

where $\tau(Q)$ is the number of divisors of Q. Deduce that

$$\sum_{\deg Q=j}{}' \tau(Q) = (j+1)q^j .$$

2. Use (5.25) and (5.26) to show that

$$\sum_{\deg Q=j}{}' \mu(Q) = \begin{cases} 1 & \text{if } j=0 \\ -q & \text{if } j=1 \\ 0 & \text{if } j \geq 2 \end{cases}$$

where $\mu(Q)$ is the Möbius function introduced in §4.4.1.

6

The polynomial 3-primes generating function

In this and the next chapter, we apply the concepts and results of Chapters 2 to 5 to obtain an asymptotic solution to the polynomial 3-primes problem. Throughout this chapter, a summation over 'P' indicates a summation over positive, irreducible polynomials. We begin with a brief look at what lies ahead.

Suppose that M is a positive polynomial in $\mathbf{A}$ of degree r. We are interested in obtaining a formula for the number $N(M)$ of representations of M as the sum of three positive irreducibles, one of degree r and the other two of degree less than r. To this end, we introduce the functions

$$F_r(t) = \sum_{\deg P = r}{}' E(Pt) \tag{6.1}$$

and

$$H_r(t) = \sum_{\deg P < r}{}' E(Pt) \tag{6.2}$$

where $t \in \mathbb{A}_k$ and where E is the additive character introduced in §4.3. If D is a fundamental domain for $\mathbb{A}_k/k$, then (cf. §§1.4 and 1.5)

$$N(M) = \int_D F_r(t) \cdot H_r(t)^2 E(-Mt)\, dt\,, \tag{6.3}$$

and our job is to approximate this integral. The Gauss sums of Chapter 3 and the technique of Fourier inversion come into play as we see that in fact

$$F_r(t) = \sum_{\deg P = r}{}' \sum_{\chi \in \widehat{W}} G(\chi, E^P)$$

where $\widehat{W}$ is the Pontryagin dual of W, the fundamental domain for $J_k^0/k^\times$ introduced in Chapter 4, and G is a global Gauss sum. (Recall from Definition 4.5 that the character E^P is defined for $t \in \mathbb{A}_k$ by $E^P(t) = E(Pt)$.)

The analysis therefore focuses on multiplicative characters, and the results of Chapter 5 become crucial. In Chapter 7, we put all the pieces together and obtain an asymptotic formula for $N(M)$.

6.1 Global Gauss sums and the summation $F_r(t)$

For each $\mathrm{v} \in \mathbf{M}_{\mathrm{fin}}$, let $d_{\mathrm{v}}u_{\mathrm{v}}$ be the Haar measure of unit total mass on $\mathbf{U}_{\mathrm{v}}$; and let $d_\infty u_\infty$ be the Haar measure of unit total mass on $\mathbf{U}_\infty^{(1)}$. If du be the product of these measures, then du is the Haar measure of unit total mass on

$$W = \left(\prod_{\mathrm{v} \in \mathbf{M}_{\mathrm{fin}}} \mathbf{U}_{\mathrm{v}} \right) \times \mathbf{U}_\infty^{(1)} .$$

For fixed $t \in \mathbb{A}_k$, the Fourier transform of E^t restricted to W at $\chi \in \widehat{W}$ is

$$G(\chi, t) = G(\chi, E^t) \stackrel{\mathrm{def}}{=} \int_W E(tu)\chi(u)\, du . \tag{6.4}$$

We call $G(\chi, t)$ a *global Gauss sum.* By definition of the product measure du and by (3.2) and (3.5),

$$G(\chi, t) = \left(\prod_{\mathrm{v} \in \mathbf{M}_{\mathrm{fin}}} G(\chi_{\mathrm{v}}, t_{\mathrm{v}}) \right) \times G^{(1)}(\chi_\infty, t_\infty) \tag{6.5}$$

where $G^{(1)}(\chi_\infty, t_\infty) = G^{(1)}(\chi_\infty, E_\infty^{t_\infty})$ and $G(\chi_{\mathrm{v}}, t_{\mathrm{v}}) = G(\chi_{\mathrm{v}}, E_{\mathrm{v}}^{t_{\mathrm{v}}})$ for each $\mathrm{v} \in \mathbf{M}_{\mathrm{fin}}$ are the local Gauss sums of Chapter 3. Applying Fourier inversion to (6.4), we get

$$E(tu) = \int_{\widehat{W}} G(\chi, t)\overline{\chi}(u)\, d\chi = \sum_{\chi \in \widehat{W}} G(\chi, t)\overline{\chi}(u)$$

since $d\chi$, the measure dual to du, is just counting measure. We will show in the next section that this Fourier series is absolutely convergent. Setting $u = 1$, we obtain

$$E(t) = \sum_{\chi \in \widehat{W}} G(\chi, t) . \tag{6.6}$$

This expansion enables us to derive information about the additive character E from information about multiplicative characters and Gauss sums.

Let $\chi \in \widehat{W}$. By the work of §5.1, we can regard χ also as a character of Dirichlet type on J_k. Before stating our first result, we must extend the concept of χ_* as follows: if χ is a character of Dirichlet type with conductor $\mathfrak{m}(\chi) = \infty^{s+1}Q$ and if M is any positive polynomial, then as in (5.7) we put

$$\chi_*(M) = \begin{cases} \prod_{v\in S}\overline{\chi}_v(M_v) & \text{if } M \text{ is prime to } Q \\ 0 & \text{otherwise} \end{cases} \tag{6.7}$$

where $S = S(\chi) \cup \{\infty\}$. We also introduce the notation $M^* = M/T^{\deg M}$. Since $\chi_\infty(T_\infty) = 1$, $\chi_\infty(M_\infty^*) = \chi_\infty(M_\infty)$.

Proposition 6.1. *Let P be a positive irreducible polynomial in* $\mathbf{A}$. *Then*

$$G(\chi, Pt) = \chi_*(P)\left(\prod_{v\neq\infty} G(\chi_v, t_v)\right)\cdot G^{(1)}(\chi_\infty, T^{\deg P}t_\infty) \tag{6.8}$$

provided only that $v_P(t_P) \geq 0$.

Proof. Let v be a finite place not corresponding to P. Then

$$\begin{aligned} G(\chi_v, Pt_v) &= \int_{\mathbf{U}_v} \chi_v(u_v)E_v(Pt_v u_v)\, d_v u_v \\ &= \int_{\mathbf{U}_v} \chi_v(P^{-1}u_v)E_v(t_v u_v)\, d_v u_v \\ &= \overline{\chi}_v(P)\cdot G(\chi_v, t_v)\,. \end{aligned}$$

For $v = \infty$, we get

$$\begin{aligned} G^{(1)}(\chi_\infty, Pt_\infty) &= \int_{\mathbf{U}_\infty^{(1)}} \chi_\infty(u_\infty)E_\infty(Pt_\infty u_\infty)\, d_\infty u_\infty \\ &= \int_{\mathbf{U}_\infty^{(1)}} \chi_\infty((P^*)^{-1}u_\infty)E((P^*)^{-1}Pt_\infty u_\infty)\, d_\infty u_\infty \\ &= \overline{\chi}_\infty(P^*)\cdot G(\chi_\infty, T^{\deg P}t_\infty)\,. \end{aligned}$$

It remains to look at the case $v = P$. Here there are two possibilities: either P is prime to the finite part of $\mathfrak{m}(\chi)$ or it is not. If it is, then χ_P is trivial. By Proposition 4.8, we have $n(E_P) = 0$ and hence

$$n(E_P^{t_P}) = n(E_P) - v_P(t_P) \leq 0$$

by (2.12) and our assumption on t_P. Similarly,

$$n(E_P^{Pt_P}) = n(E_P) - v_P(t_P) - v_P(P) < 0\,.$$

We conclude from Theorem 3.3 therefore that

$$G(\chi_P, Pt_P) = \overline{\chi}_P(P)\cdot G(\chi_P, t_P)$$

because both sides have the value one. On the other hand, if P divides $\mathfrak{m}(\chi)$, then the right hand side of (6.8) is zero by (6.7). However, $m(\chi_P) \geq 1$

whereas $n(E_P^{Pt_P}) < 0$, so by Theorem 3.4 $G(\chi_P, Pt_P)$ is also zero. This completes the proof. ■

We can now write down a Fourier expansion for the function $F_r(t)$ defined by (6.1). Applying (6.6) and Proposition 6.1 and recalling (5.10) from Chapter 5, we find that

$$F_r(t) = \sum_{\chi \in \widehat{W}} \pi_r(\chi) \cdot \prod_{\mathrm{v} \neq \infty} G(\chi_{\mathrm{v}}, t_{\mathrm{v}}) \cdot G^{(1)}(\chi_\infty, T^r t_\infty) \qquad (6.9)$$

provided that $\mathrm{v}_P(t_P) \geq 0$ for all P with $\deg P = r$. Thus, $F_r(t)$ is a summation over $\widehat{W}$, and we are interested in splitting off the term with χ trivial, for this will be the 'major term' in the analysis. For n a positive integer, let

$$\psi(n) = \#\{P : \deg(P) = n\}, \qquad (6.10)$$

so that for $\chi = \chi_0$, we have

$$\pi_r(\chi_0) = \sum_{\deg P = r}{}' \chi_0(P) = \psi(r).$$

If $\mathrm{v} \neq \infty$, then since $n(E_{\mathrm{v}}^{t_{\mathrm{v}}}) = -\mathrm{v}(t_{\mathrm{v}})$, Theorem 3.3 tells us that for $\chi = \chi_0$

$$G(\chi_{\mathrm{v}}, t_{\mathrm{v}}) = \begin{cases} 1 & \text{if } \mathrm{v}(t_{\mathrm{v}}) \geq 0 \\ \frac{1}{1-q_{\mathrm{v}}} & \text{if } \mathrm{v}(t_{\mathrm{v}}) = -1 \\ 0 & \text{if } \mathrm{v}(t_{\mathrm{v}}) < -1 \end{cases}$$

where q_{v} is the order of the residue class field $\mathbf{K}(\mathrm{v})$. For $\mathrm{v} = \infty$, we have $n(E_\infty) = 2$ by Proposition 4.8, and so $n(E_\infty^{T^r t_\infty}) = 2 + r - \mathrm{v}_\infty(t_\infty)$, giving us that $n(E_\infty^{T^r t_\infty}) \leq 1$ if and only if $\mathrm{v}_\infty(t_\infty) > r$. Hence Theorem 3.5 implies that

$$G^{(1)}(\chi_\infty, T^r t_\infty) = \begin{cases} E_\infty(T^r t_\infty) & \text{if } \mathrm{v}_\infty(t_\infty) > r \\ 0 & \text{otherwise.} \end{cases}$$

We have thus proved the following.

Proposition 6.2. *Suppose that* $t \in \mathbb{A}_k$ *satisfies the following conditions:* (1) $\mathrm{v}(t_{\mathrm{v}}) \geq -1$ *for all* $\mathrm{v} \neq \infty$, (2) $\mathrm{v}_P(t_P) \geq 0$ *for all* P *with* $\deg P = r$, *and* (3) $\mathrm{v}_\infty(t_\infty) > r$. *Then in the Fourier expansion* (6.9), *the term corresponding to* $\chi = \chi_0$ *is*

$$\psi(r) \cdot E_\infty(T^r t_\infty) \cdot \prod_{\substack{\mathrm{v} \\ \mathrm{v}(t_{\mathrm{v}}) = -1}} \left(\frac{1}{1 - q_{\mathrm{v}}} \right)$$

where ψ *is defined by* (6.10). *Moreover, if either condition* (1) *or condition* (3) *fails, this term is zero.*

This result gives us the value of the 'χ-trivial' term of (6.9), and our hope is that the magnitude of this term exceeds that of the sum of the absolute values of the remaining terms, at least for sufficiently large values of q and r. Our next task is to obtain an upper bound for the absolute value of the remaining terms by evaluating $\int_{\widehat{W}} |G(\chi,t)|\,d\chi$ for any fixed $t \in \mathbb{A}_k$.

6.2 An estimate for $\sum_{\chi\in\widehat{W}}|G(\chi,t)|\,d\chi$

We begin this section with some definitions. If $t \in \mathbb{A}_k$, then for $\mathrm{v} \neq \infty$ we set

$$\mathfrak{n}_{\mathrm{v}}(t_{\mathrm{v}}) \stackrel{\mathrm{def}}{=} \mathfrak{p}_{\mathrm{v}}^{\max\{0,-\mathrm{v}(t_{\mathrm{v}})\}} \tag{6.11}$$

where $\mathfrak{p}_{\mathrm{v}}$ is the prime divisor associated with the place v (see §4.3). We observe that $\mathfrak{n}_{\mathrm{v}}(t_{\mathrm{v}})$ is an integral divisor which measures how far t_{v} is from being v-integral. For $\mathrm{v} = \infty$, we set

$$\mathfrak{n}_{\infty}(t_{\infty}) \stackrel{\mathrm{def}}{=} \begin{cases} 1 & \text{if } \mathrm{v}_{\infty}(t_{\infty}) > 0 \\ \infty^{2-\mathrm{v}_{\infty}(t_{\infty})} & \text{if } \mathrm{v}_{\infty}(t_{\infty}) \leq 0\,, \end{cases} \tag{6.12}$$

and we put

$$\mathfrak{n}(t) \stackrel{\mathrm{def}}{=} \prod_{\mathrm{v}\in\mathbf{M}} \mathfrak{n}_{\mathrm{v}}(t_{\mathrm{v}}) \tag{6.13}$$

and

$$\tilde{\mathfrak{n}}(t) \stackrel{\mathrm{def}}{=} \prod_{\mathrm{v}\in\mathbf{M}_{\mathrm{fin}}} \mathfrak{n}_{\mathrm{v}}(t_{\mathrm{v}})\,. \tag{6.14}$$

Both these definitions make sense because every $t \in \mathbb{A}_k$ is v-integral almost everywhere. The divisor $\mathfrak{n}(t)$ is called the *conductor* of t.

We recall the definition of the norm $|\mathfrak{d}|$ of a divisor $\mathfrak{d}$ from (4.44). If $\mathfrak{d}$ is an integral divisor, we put

$$\Phi(\mathfrak{d}) \stackrel{\mathrm{def}}{=} \prod_{\substack{\mathfrak{p}_{\mathrm{v}}|\mathfrak{d} \\ \mathrm{v}\neq\infty}} \left(|\mathfrak{p}_{\mathrm{v}}|^{s_{\mathrm{v}}} - |\mathfrak{p}_{\mathrm{v}}|^{s_{\mathrm{v}}-1}\right) \cdot \begin{cases} |\infty|^{s_{\mathrm{v}}-1} & \text{if } \infty|\mathfrak{d} \\ 1 & \text{otherwise} \end{cases} \tag{6.15}$$

where for each $\mathrm{v} \in \mathbf{M}$, s_{v} is the highest power of $\mathfrak{p}_{\mathrm{v}}$ dividing $\mathfrak{d}$. The function Φ is an Euler function on divisors. We will also need a Möbius function on divisors. Suppose that $\mathfrak{d}$ is integral and relatively prime to ∞. Then we put

$$\mu(\mathfrak{d}) \stackrel{\mathrm{def}}{=} \begin{cases} 0 & \text{if } \mathfrak{d} \text{ is not square-free} \\ (-1)^n & \text{otherwise} \end{cases} \tag{6.16}$$

where n equals the number of prime divisors of $\mathfrak{d}$. When $\mathfrak{d}$ is the finite part of the divisor $\partial(Q)$ of a positive polynomial $Q \in \mathbf{A}$, then $\Phi(\mathfrak{d}) = \Phi(Q)$ and $\mu(\mathfrak{d}) = \mu(Q)$ as defined in (4.30) and (4.29).

Given $t \in \mathbb{A}_k$, we write $\mathfrak{n}(t) = \mathfrak{n}_1(t) \cdot \mathfrak{n}_2(t)$, where $\mathfrak{n}_1$ and $\mathfrak{n}_2$ are relatively prime, $\mathfrak{n}_1$ is square-free, and $\mathfrak{n}_2$ is *square-full*, i.e. $\mathfrak{p}_v | \mathfrak{n}_2$ implies that $\mathfrak{p}_v^2 | \mathfrak{n}_2$. Let S_1 (resp. S_2) be the set of places v such that $\mathfrak{p}_v | \mathfrak{n}_1$ (resp. $\mathfrak{p}_v | \mathfrak{n}_2$), and let S_3 be complement of $S_1 \cup S_2$ in $\mathbf{M}$. We may then decompose W as follows:

$$W = W_1 \times W_2 \times W_3$$

where for $i = 1, 2$, and 3,

$$W_i = \prod_{v \in S_i} \mathbf{U}_v^*$$

and where by $\mathbf{U}_v^*$ we mean $\mathbf{U}_v$ if $v \neq \infty$ and $\mathbf{U}_\infty^{(1)}$ if $v = \infty$. We note from (6.12) that ∞ never divides $\mathfrak{n}_1(t)$.

Now t has a unique decomposition $t = t_1 + t_2 + t_3$, where for $i = 1, 2$, and 3, $(t_i)_v = t_v$ if $v \in S_i$ and is zero otherwise. Since $\widehat{W}$ is naturally isomorphic to $\widehat{W}_1 \times \widehat{W}_2 \times \widehat{W}_3$, we have

$$\int_{\widehat{W}} |G(\chi, t)| \, d\chi = \prod_{i=1}^{3} \int_{\widehat{W}_i} |G(\chi_i, t_i)| \, d\chi_i \tag{6.17}$$

because $G(\chi, t) = G(\chi_1, t_1) G(\chi_2, t_2) G(\chi_3, t_3)$ by (6.5). We may now state the two main results of this section.

Theorem 6.3. *We have*

$$\int_{\widehat{W}} |G(\chi, t)| \, d\chi = |\mathfrak{n}_1(t)|^{\frac{1}{2}} \cdot \frac{\Phi(\mathfrak{n}_2(t))}{|\mathfrak{n}_2(t)|^{\frac{1}{2}}} \prod_{v \in S_1} \left(1 - \frac{1 - |\mathfrak{p}_v|^{-\frac{1}{2}}}{|\mathfrak{p}_v|^{-1}} \right).$$

Proof. The proof follows from (6.17) by Lemmas 6.5, 6.6, and 6.8 below. ∎

Corollary 6.4. *We have*

$$\int_{\widehat{W}} |G(\chi, t)| \, d\chi < \begin{cases} (\Phi(\mathfrak{n}(t))/q)^{\frac{1}{2}} & \textit{if } \infty | \mathfrak{n}(t) \\ \Phi(\mathfrak{n}(t))^{\frac{1}{2}} & \textit{otherwise.} \end{cases}$$

Proof. First we observe that the $\mathfrak{n}_2$ part of the right hand side of Theorem 6.3 is less than $\Phi(\mathfrak{n}_2(t))^{\frac{1}{2}}$ in the case that $\infty \nmid \mathfrak{n}(t)$ and is less than

$\Phi(\mathfrak{n}_2(t))^{\frac{1}{2}}/\sqrt{q}$ if $\infty|\mathfrak{n}(t)$ since then $|\mathfrak{n}_\infty(t_\infty)|/\Phi(\mathfrak{n}_\infty(t_\infty)) = |\infty| = q$. For the $\mathfrak{n}_1$ part, since each $\mathfrak{p}_{\mathrm{v}}|\mathfrak{n}_1(t)$ appears linearly, it suffices to prove that

$$|\mathfrak{p}_{\mathrm{v}}|^{\frac{1}{2}}\left(1-\frac{1-|\mathfrak{p}_{\mathrm{v}}|^{-\frac{1}{2}}}{|\mathfrak{p}_{\mathrm{v}}|-1}\right)\le \Phi(\mathfrak{p}_{\mathrm{v}})^{\frac{1}{2}} = (|\mathfrak{p}_{\mathrm{v}}|-1)^{\frac{1}{2}}\,.$$

The reader can easily verify that the inequality

$$x^{\frac{1}{2}}\left(1-\frac{1-x^{-\frac{1}{2}}}{x-1}\right)\le (x-1)^{\frac{1}{2}}$$

is valid for all $x\ge 2$. ∎

In order to prove Theorem 6.3, we will separately evaluate the three integrals in (6.17).

Lemma 6.5. *We have*

$$\int_{\widehat{W}_3}|G(\chi_3,t_3)|\,d\chi_3 = \sum_{\chi_3\in\widehat{W}_3}|G(\chi_3,t_3)|\,d\chi_3 = 1\,.$$

Proof. Let $\mathrm{v}\in S_3$. If $\mathrm{v}\ne\infty$, then $n(E_{\mathrm{v}}^{t_{\mathrm{v}}}) = -\mathrm{v}(t_{\mathrm{v}})\le 0$ since $\mathfrak{p}_{\mathrm{v}}\not|\,\mathfrak{n}(t)$. If $\mathrm{v}=\infty$, then we have $n(E_\infty^{t_\infty}) = 2-\mathrm{v}_\infty(t_\infty)\le 1$ by (6.12).

If χ_3 is non-trivial, then there is a $\mathrm{v}\in S_3$ such that the restriction χ_{v} of χ_3 to $\mathbf{U}_{\mathrm{v}}^*$ is non-trivial. If $\mathrm{v}\ne\infty$, then the local Gauss sum $G_{\mathrm{v}}(\chi_{\mathrm{v}},E_{\mathrm{v}}^{t_{\mathrm{v}}})=0$ by Theorem 3.4. If $\mathrm{v}=\infty$, then the modified Gauss sum $G_\infty^{(1)}(\chi_{\mathrm{v}},E_\infty^{t_\infty})=0$ by Theorem 3.6. In either case, $G(\chi_3,t_3)=0$ by the product decomposition (6.5). However, if χ_3 is the trivial character, then $G(\chi_3,t_3)=1$ by (6.5) and Theorems 3.3 and 3.5. ∎

Lemma 6.6. *We have*

$$\int_{\widehat{W}_1}|G(\chi_1,t_1)|\,d\chi_1 = |\mathfrak{n}_1(t)|^{\frac{1}{2}}\prod_{\mathrm{v}\in S_1}\left(1-\frac{1-|\mathfrak{p}_{\mathrm{v}}|^{-\frac{1}{2}}}{|\mathfrak{p}_{\mathrm{v}}|-1}\right).$$

Proof. Since S_1 is finite, the integral over $\widehat{W}_1$ is a finite product of local integrals by (6.5). Let $\mathrm{v}\in S_1$. Then $\mathrm{v}\ne\infty$ and $\mathrm{v}(t_{\mathrm{v}})=-1$, so that

$n(E_{\mathrm{v}}^{t_{\mathrm{v}}}) = 1$. If the restriction χ_{v} of χ_1 to $\mathbf{U}_{\mathrm{v}}$ is trivial, then Theorem 3.3 applies and gives us

$$|G_{\mathrm{v}}(\chi_{\mathrm{v}}, t_{\mathrm{v}})| = \frac{1}{q_{\mathrm{v}} - 1} = \frac{1}{|\mathfrak{p}_{\mathrm{v}}| - 1}.$$

For χ_{v} non-trivial, Theorem 3.4 shows that the local Gauss sum is zero if $m(\chi_{\mathrm{v}}) \neq 1$ and equals

$$\frac{1}{q_{\mathrm{v}} - 1} q_{\mathrm{v}}^{\frac{1}{2}} = \frac{|\mathfrak{p}_{\mathrm{v}}|^{\frac{1}{2}}}{|\mathfrak{p}_{\mathrm{v}}| - 1}$$

otherwise. Now the number of characters of $\mathbf{U}_{\mathrm{v}}$ which satisfy $m(\chi_{\mathrm{v}}) = 1$ is $\#(\widehat{\mathbf{U}}_{\mathrm{v}}/\widehat{\mathbf{U}}_{\mathrm{v}}^{(1)}) - 1 = (q_{\mathrm{v}} - 1) - 1 = |\mathfrak{p}_{\mathrm{v}}| - 2$. Hence the integral over $\widehat{\mathbf{U}}_{\mathrm{v}}$ equals

$$\begin{aligned}\sum_{\chi_{\mathrm{v}} \in \widehat{\mathbf{U}}_{\mathrm{v}}} |G_{\mathrm{v}}(\chi_{\mathrm{v}}, t_{\mathrm{v}})| &= \frac{1}{|\mathfrak{p}_{\mathrm{v}}| - 1} + (|\mathfrak{p}_{\mathrm{v}}| - 2)\frac{|\mathfrak{p}_{\mathrm{v}}|^{\frac{1}{2}}}{|\mathfrak{p}_{\mathrm{v}}| - 1} \\ &= |\mathfrak{p}_{\mathrm{v}}|^{\frac{1}{2}} \left(1 - \frac{1 - |\mathfrak{p}_{\mathrm{v}}|^{-\frac{1}{2}}}{|\mathfrak{p}_{\mathrm{v}}| - 1}\right).\end{aligned}$$

Multiplying over all v involved in W_1, we obtain the result. ■

For W_2, we will need the following.

Lemma 6.7. *The number of characters χ_{v} in $\mathbf{U}_{\mathrm{v}}^*$ which satisfy $\mathfrak{m}_{\mathrm{v}}(\chi_{\mathrm{v}}) = \mathfrak{n}_{\mathrm{v}}(t_{\mathrm{v}}) = n$ and are n-cancellative if* $\mathrm{v} = \infty$ *is exactly*

$$\frac{\Phi(\mathfrak{n}_{\mathrm{v}}(t_{\mathrm{v}}))^2}{|\mathfrak{n}_{\mathrm{v}}(t_{\mathrm{v}})|}.$$

Proof. Suppose first that $\mathrm{v} \neq \infty$. The number of characters of $\mathbf{U}_{\mathrm{v}}$ which are trivial on $\mathbf{U}_{\mathrm{v}}^{(n)}$ but not on $\mathbf{U}_{\mathrm{v}}^{(n-1)}$ is exactly

$$\begin{aligned}\#\left(\widehat{\mathbf{U}}_{\mathrm{v}}/\widehat{\mathbf{U}}_{\mathrm{v}}^{(n)}\right) - \#\left(\widehat{\mathbf{U}}_{\mathrm{v}}/\widehat{\mathbf{U}}_{\mathrm{v}}^{(n-1)}\right) &= q_{\mathrm{v}}^{n-1}(q_{\mathrm{v}} - 1) - q_{\mathrm{v}}^{n-2}(q_{\mathrm{v}} - 1) \\ &= (q_{\mathrm{v}} - 1)^2 q_{\mathrm{v}}^{n-2} \\ &= (q_{\mathrm{v}}^n - q_{\mathrm{v}}^{n-1})^2 q_{\mathrm{v}}^{2-2n} q_{\mathrm{v}}^{n-2} \\ &= \Phi(\mathfrak{n}_{\mathrm{v}}(t_{\mathrm{v}}))^2 / |\mathfrak{n}_{\mathrm{v}}(t_{\mathrm{v}})|\end{aligned}$$

(cf. Exercise 3.1(1)).

Now let $v=\infty$. The characters $E_\infty^{t_\infty}$ and χ_∞ are n-cancellative if and only if χ_∞ is trivial on $\mathbf{U}_\infty^{(n)}$ and restricts to $E_\infty^{t_\infty}$ on $\mathbf{U}_\infty^{(n-1)}$. Therefore the number of characters in this case is

$$\#\left(\widehat{\mathbf{U}}_\infty^{(1)}/\widehat{\mathbf{U}}_\infty^{(n-1)}\right)=q^{n-2}=q^{2n-2}/q^n=\Phi(\mathfrak{n}_\infty(t_\infty))^2/|\mathfrak{n}_\infty(t_\infty)|.$$

■

We may now evaluate the remaining part of the integral of Theorem 6.3.

Lemma 6.8. *We have*

$$\int_{\widehat{W}_2}|G(\chi_2,t_2)|\,d\chi_2=\frac{\Phi(\mathfrak{n}_2(t))}{|\mathfrak{n}_2(t)|^{\frac12}}.$$

Proof. Like $\widehat{W}_1$, $\widehat{W}_2$ can be viewed as a finite product. Therefore we may write

$$\int_{\widehat{W}_2}|G(\chi_2,t_2)|\,d\chi_2=\prod_{v\in S_2}\int_{\widehat{\mathbf{U}}_v^*}|G^*(\chi_v,t_v)|\,d\chi_v$$

where $G^*(\chi_v,t_v)=G(\chi_v,t_v)$ if $v\neq\infty$ and $G^*(\chi_\infty,t_\infty)=G^{(1)}(\chi_\infty,t_\infty)$. Suppose first that $v\neq\infty$. If χ_v is trivial, then Theorem 3.3 guarantees that $|G(\chi_v,t_v)|=0$ since $n(E_v^{t_v})=-v(t_v)\geq 2$ by definition of S_2. If χ_v is non-trivial, Theorem 3.4 implies that

$$|G(\chi_v,t_v)|=\frac{1}{q_v^n-q_v^{n-1}}\cdot q_v^{n/2}=|\mathfrak{n}_v(t_v)|^{\frac12}/\Phi(\mathfrak{n}_v(t_v))$$

provided that $n=n(E_v^{t_v})=m(\chi_v)$; but $G(\chi_v,t_v)=0$ if $m(\chi_v)\neq m(\chi_v)$.

Now suppose that $v=\infty$. Once again χ_∞ trivial implies (by Theorem 3.5) that $G_\infty^{(1)}(\chi_\infty,t_\infty)=0$. If χ_∞ is non-trivial, Theorems 3.6 and 3.7 tell us that we get zero unless χ_∞ and E^{t_∞} are n-cancellative as defined in (3.7), in which case we get

$$\frac{1}{q^{n-1}}q^{n/2}=\frac{|\mathfrak{n}_\infty(t_\infty)|^{\frac12}}{\Phi(\mathfrak{n}_\infty(t_\infty))}.$$

We have shown then that

$$\sum_{\chi_v\in\widehat{\mathbf{U}}_v^*}|G^*(\chi_v,t_v)|=N\cdot\frac{|\mathfrak{n}_v(t_v)|^{\frac12}}{\Phi(\mathfrak{n}_v(t_v))}$$

where N is the number of $\chi_v\in\widehat{\mathbf{U}}_v^*$ which satisfy $\mathfrak{m}_v(\chi_v)=\mathfrak{n}_v(t_v)=n$, and in addition χ_∞ and $E_\infty^{t_\infty}$ are n-cancellative. But Lemma 6.7 evaluates N, and the global result now follows from the multiplicativity of the norm and the Φ function. ■

6.3 An asymptotic expression for $F_r(t)$ and the fundamental domains

We are now in a position to write down an asymptotic expression for the crucial summation $F_r(t)$ defined in (6.1). This expression will then guide us in the formation of fundamental domains for $\mathbb{A}_k/k$ which will minimize the error in the analysis to be carried out in Chapter 7. The main result of Chapter 5 (Theorem 5.7) will also come into play at this point.

We introduce two additional notations: first if $t \in \mathbb{A}_k$, then we define $t^{(r)}$ by

$$t_{\mathrm{v}}^{(r)} \stackrel{\text{def}}{=} \begin{cases} t_{\mathrm{v}} & \text{if } \mathrm{v} \neq \infty \\ T^r t_\infty & \text{if } \mathrm{v} = \infty . \end{cases} \tag{6.18}$$

We observe that since $\mathrm{v}_\infty(T^r t_\infty) = -r + \mathrm{v}_\infty(t_\infty)$, we have

$$\mathfrak{n}_\infty(t_\infty^{(r)}) = \begin{cases} 1 & \text{if } \mathrm{v}_\infty(t_\infty) > r \\ \infty^{2+r-\mathrm{v}_\infty(t_\infty)} & \text{if } \mathrm{v}_\infty(t_\infty) \leq r \end{cases} \tag{6.19}$$

by (6.12). We also need the characteristic function of the complement of $\pi_\infty \mathcal{O}_\infty$ in k_∞:

$$\delta(t_\infty) \stackrel{\text{def}}{=} \begin{cases} 1 & \text{if } \mathrm{v}_\infty(t_\infty) \leq 0 \\ 0 & \text{if } \mathrm{v}_\infty(t_\infty) > 0 . \end{cases} \tag{6.20}$$

We note from (6.12) and (6.15) that

$$\frac{|\mathfrak{n}_\infty(t_\infty)|}{\Phi(\mathfrak{n}_\infty(t_\infty))} = q^{\delta(t_\infty)} . \tag{6.21}$$

Theorem 6.9. *Suppose that* $\mathrm{v}_P(t_P) \geq 0$ *for all* P *with* $\deg P = r$ *and that* $\deg \mathfrak{n}(t) = O(r)$. *Then*

$$F_r(t) = \Upsilon_r(t) + O\left(q^{\frac{r+\deg \mathfrak{n}(t^{(r)})-2\delta(T^r t_\infty)}{2}}\right) \tag{6.22}$$

where

$$\Upsilon_r(t) = \psi(r)\frac{\mu(\mathfrak{n}(t))}{\Phi(\mathfrak{n}(t))}(1-\delta(T^r t_\infty)) \cdot E_\infty(T^r t_\infty) \tag{6.23}$$

and where $F_r(t)$, $\psi(r)$, $\mathfrak{n}(t)$, Φ, μ, $t^{(r)}$, *and* δ *are defined in* (6.1), (6.10), (6.13), (6.15), (6.16), (6.18), *and* (6.20) *respectively.*

Proof. We first look at the term corresponding to χ trivial in the Fourier expansion (6.9). Proposition 6.2 gives an expression for this term under three conditions, the second of which is assumed here. If there exists a place v for which $\mathrm{v}(t_{\mathrm{v}}) < -1$ then $\mu(\mathfrak{n}(t)) = 0$ by definition, and if $\mathrm{v}_\infty(t_\infty) \leq r$

then $\delta(T^r t_\infty) = 1$, so in either case the main term in the theorem is zero, matching the result in Proposition 6.2. We assume then that $\mathrm{v}(t_\mathrm{v}) \geq -1$ for all $\mathrm{v} \in \mathbf{M}_{\mathrm{fin}}$. Since the non-trivial factors in the conductor $\mathfrak{n}(t)$ all satisfy $\mathrm{v}(t_\mathrm{v}) = -1$ for such v, $q_\mathrm{v} - 1 = \Phi(\mathfrak{n}_\mathrm{v}(t_\mathrm{v}))$, so that

$$\frac{1}{1-q_\mathrm{v}} = \frac{\mu(\mathfrak{n}_\mathrm{v}(t_\mathrm{v}))}{\Phi(\mathfrak{n}_\mathrm{v}(t_\mathrm{v}))}.$$

Finally for $\mathrm{v} = \infty$, $\mathrm{v}_\infty(t_\infty) > r$ gives $\mathfrak{n}_\infty(t_\infty) = 1$. By multiplicativity of μ and Φ, we conclude that the χ_0-term of (6.9) is just $\Upsilon_r(t)$.

We now turn to the error term. By (6.9) and (6.18), that term is

$$\sum_{\chi \neq \chi_0} \pi_r(\chi) G(\chi, t^{(r)})$$

where $\pi_r(\chi)$ is defined by (5.10). By Theorems 3.4 and 3.6, any summand in (6.9) for which $\mathfrak{m}(\chi) \nmid \mathfrak{n}(t^{(r)}) \cdot \infty$ is zero (cf. Lemma 7.3). Applying Theorem 5.7 and the hypothesis that $\deg \mathfrak{n}(t) = O(r)$, we obtain

$$\left| \sum_{\chi \neq \chi_0} \pi_r(\chi) G(\chi, t^{(r)}) \right| \leq C q^{r/2} \sum_{\text{all } \chi} \left| G(\chi, t^{(r)}) \right| \tag{6.24}$$

for some real constant C. We now apply Theorem 6.3 to the latter sum. Observing that each term in the final product there is less than one, and also that $\Phi(\tilde{\mathfrak{n}}_2(t)) \leq |\tilde{\mathfrak{n}}_2(t)|$ (see (6.14)), we obtain

$$\begin{aligned} \sum_{\chi} |G(\chi, t^{(r)})| &\leq |\mathfrak{n}_1(t)|^{\frac{1}{2}} \frac{\Phi(\mathfrak{n}_2(t))}{|\mathfrak{n}_2(t)|^{\frac{1}{2}}} \leq |\tilde{\mathfrak{n}}(t)|^{\frac{1}{2}} \frac{\Phi(\mathfrak{n}_\infty(T^r t_\infty))}{|\mathfrak{n}_\infty(T^r t_\infty)|^{\frac{1}{2}}} \\ &= |\mathfrak{n}(t^{(r)})|^{\frac{1}{2}} \frac{\Phi(\mathfrak{n}_\infty(T^r t_\infty))}{|\mathfrak{n}_\infty(T^r t_\infty)|} = |\mathfrak{n}(t^{(r)})|^{\frac{1}{2}} q^{-\delta(T^r t_\infty)} \end{aligned}$$

by (6.21). Combining this with (6.24), we see that the absolute value of the sum of the terms in (6.9) with $\chi \neq \chi_0$ is bounded above by

$$C q^{r/2} \left(q^{\frac{\deg \mathfrak{n}(t^{(r)}) - 2\delta(T^r t_\infty)}{2}} \right)$$

as required. ■

We note for future use a fact which is clear from Theorems 3.3 and 3.5 and the first part of the proof of Theorem 6.9.

Corollary 6.10. *We have*

$$G(\chi_0, t^{(r)}) = \begin{cases} \frac{\mu(\mathfrak{n}(t^r))}{\Phi(\mathfrak{n}(t^r))} E_\infty(T^r t_\infty) & \text{if } \mathrm{v}_\infty(t_\infty) > r \\ 0 & \text{if } \mathrm{v}_\infty(t_\infty) \le r\,. \end{cases}$$

Although we require a more precise estimate for $F_r(t)$ in the next chapter, Theorem 6.9 tells us the magnitude of the error term. We shall now let this result guide us in the formation of fundamental domains for $\mathbb{A}_k/k$ which minimize the effect of this error. To this end, let $\epsilon \ge 0$ be a real number and let Q be a positive polynomial in $\mathbf{A}$ of degree j. We put

$$\mathcal{T}_Q = \prod_{P^s \| Q} (\pi_P^{-s}\mathcal{O}_P - \pi_P^{-s+1}\mathcal{O}_P) \tag{6.25}$$

where $P^s\|Q$ means that s is the highest power of P dividing Q. The set $\mathcal{T}_Q$ is called a *Q-toroid.* Now let

$$D_Q^\epsilon = \left(\prod_{P\nmid Q} \mathcal{O}_P\right) \times \mathcal{T}_Q \times \pi_\infty^{[\epsilon]+j+1}\mathcal{O}_\infty \tag{6.26}$$

where $[\epsilon]$ is the greatest integer in ϵ. Finally, let

$$D^\epsilon = \bigcup_{\deg Q \le \epsilon} D_Q^\epsilon\,. \tag{6.27}$$

Proposition 6.11. *For each $\epsilon \ge 0$, D^ϵ is a fundamental domain for $\mathbb{A}_k/k$.*

Proof. We first compute the dt-measure of D^ϵ. Since $meas(\pi_\mathrm{v}^{-s}\mathcal{O}_\mathrm{v}) = q_\mathrm{v}^s$ by (2.14), we have

$$meas(\mathcal{T}_Q) = \prod_{P^s\|Q} (q_P^s - q_P^{s-1}) = \Phi(Q)\,.$$

Also $meas(\pi_\infty^{[\epsilon]+j+1}\mathcal{O}_\infty) = q^{-[\epsilon]-j}$. Now since the union which makes up D^ϵ is clearly disjoint, we compute

$$meas(D^\epsilon) = \sum_{\deg Q \le \epsilon} \Phi(Q) q^{-[\epsilon]-j} = q^{-[\epsilon]} \sum_{j=0}^{[\epsilon]} q^{-j} \sum_{\deg Q = j}{}' \Phi(Q)\,.$$

Now Carlitz (1932) has shown that

$$\sum_{\deg Q=j}{}' \Phi(Q) = \begin{cases} 1 & \text{if } j=0 \\ q^{2j}-q^{2j-1} & \text{if } j \geq 1 . \end{cases} \tag{6.28}$$

Therefore

$$meas(D^{\epsilon}) = q^{-[\epsilon]}\left(1+\sum_{j=1}^{[\epsilon]} q^{-j}(q^{2j}-q^{2j-1})\right) = 1 .$$

By its definition, D^{ϵ} is both compact and open, so it remains to show that every element of $\mathbb{A}_k/k$ has a unique representative in D^{ϵ}. Suppose that $t, t' \in D^{\epsilon}$ and that $t - t' = A/B$ with $A, B \in \mathbf{A}$ and relatively prime. We will show that $t = t'$. Let $t \in D^{\epsilon}_{Q_1}$ and $t' \in D^{\epsilon}_{Q_2}$ where $\deg Q_1 = j_1$ and $\deg Q_2 = j_2$. Since for every $\mathrm{v} \neq \infty$, $\mathrm{v}(A/B) = \mathrm{v}(t_{\mathrm{v}} - t'_{\mathrm{v}}) \geq \min\{\mathrm{v}(t_{\mathrm{v}}), \mathrm{v}(t'_{\mathrm{v}})\}$, the highest power of P_{v} dividing B is less than or equal to the higher of the two powers of P_{v} dividing Q_1 and Q_2. Hence $B|Q_1Q_2$, so $\deg B \leq j_1 + j_2$. Let us assume that $\min\{j_1, j_2\} = j_1$. Then

$$\begin{aligned} \deg B - \deg A &= \mathrm{v}_{\infty}(A/B) \\ &\geq \min\{\mathrm{v}_{\infty}(t_{\infty}), \mathrm{v}_{\infty}(t'_{\infty})\} \\ &\geq [\epsilon] + j_1 + 1 > [\epsilon] + j_1 . \end{aligned}$$

If $A \neq 0$, we would have $j_1 + j_2 \geq \deg B \geq \deg B - \deg A > [\epsilon] + j_1$, implying that $j_2 > [\epsilon]$, which is a contradiction. Hence $A = 0$ and we have proved uniqueness.

Consider now the projection of D^{ϵ} into $\mathbb{A}_k/k$, which we have just shown to be one-to-one. Since $meas(D^{\epsilon}) = 1$, the measure of the image must also be one. Moreover, the image of D^{ϵ} is compact, so that its complement in $\mathbb{A}_k/k$ is open and of measure zero and hence empty. ■

We finish this section by looking at three examples of ϵ-fundamental domains.

Example. Let $\epsilon = 0$. Then

$$D^0 = \bigcup_{\deg Q \leq 0} D^0_Q = \left(\prod_{\mathrm{v}\neq\infty} \mathcal{O}_{\mathrm{v}}\right) \times \pi_{\infty}\mathcal{O}_{\infty} = D_0$$

as defined in (4.2).

Example. Let $\epsilon = r/2$ and suppose that r is odd. Let us call this domain D^{odd}. We claim that

$$D^{\text{odd}} = \left\{ t \in \mathbb{A}_k : \deg \mathfrak{n}(t^{(r)}) - 2\delta(T^r t_\infty) \le \frac{r-1}{2} \right\}. \tag{6.29}$$

To prove (6.29), let $t \in \mathbb{A}_k$ and suppose that $\deg \tilde{\mathfrak{n}}(t) = j$. We have

$$\deg \mathfrak{n}(t^{(r)}) = \begin{cases} j + 2 - \mathrm{v}_\infty(T^r t_\infty) & \text{if } \delta(T^r t_\infty) = 1 \\ j & \text{if } \delta(T^r t_\infty) = 0 \end{cases}$$

by (6.12) and hence

$$\deg \mathfrak{n}(t^{(r)}) - 2\delta(T^r t_\infty) = \begin{cases} j + r - \mathrm{v}_\infty(t_\infty) & \text{if } \mathrm{v}_\infty(t_\infty) \le r \\ j & \text{if } \mathrm{v}_\infty(t_\infty) > r \end{cases} \tag{6.30}$$

by (6.20).

Suppose that $t \in D^{\text{odd}}$, so $t \in D^{\text{odd}}_Q$ for some Q. By definition of D^{odd}_Q, we have $Q = \tilde{\mathfrak{n}}(t)$, and so $\deg \tilde{\mathfrak{n}}(t) = j \le [\epsilon] = \frac{r-1}{2}$. Also by definition of D^{odd}_Q, $\mathrm{v}_\infty(t_\infty) \ge j + \frac{r-1}{2} + 1$. Hence by (6.30)

$$\begin{aligned} \deg \mathfrak{n}(t^{(r)}) - 2\delta(T^r t_\infty) &\le \begin{cases} j + r - (j + \frac{r-1}{2} + 1) & \text{if } \mathrm{v}_\infty(t_\infty) \le r \\ j & \text{if } \mathrm{v}_\infty(t_\infty) > r \end{cases} \\ &\le \frac{r-1}{2} \end{aligned}$$

in either case, as desired.

On the other hand, suppose $t \in \mathbb{A}_k$ belongs to the right hand side of (6.29). Then by (6.30) we have $\deg \tilde{\mathfrak{n}}(t) = j \le (r-1)/2$ and

$$j + r - \mathrm{v}_\infty(t_\infty) \le \frac{r-1}{2} \Rightarrow \mathrm{v}_\infty(t_\infty) \ge j + \frac{r-1}{2} + 1 \Rightarrow t \in D^{\text{odd}}$$

as desired.

Example. Let $\epsilon = r/2$ where r is even, and let $D^{\text{even}} = D^\epsilon$. An argument similar to the one just given shows that

$$D^{\text{even}} = \left\{ t \in \mathbb{A}_k : \deg \mathfrak{n}(t^{(r)}) - \delta(T^r t_\infty) \le r/2 \right\}. \tag{6.31}$$

Note the absence of the multiplier '2' in this equation. In the next chapter, D^{odd} and D^{even} will be the fundamental domains for $\mathbb{A}_k/k$ used to solve the polynomial 3-primes problem.

Exercises 6.3

1. Use (5.25) and (5.26) to show that

$$Z_\Phi(s) = \prod_P \left(\frac{1 - |P|^{-s}}{1 - |P|^{1-s}} \right) = \frac{1 - q^{1-s}}{1 - q^{2-s}}.$$

   Deduce (6.28) from this identity.

7

The polynomial 3-primes problem: an asymptotic solution

Let M be a positive polynomial in $\mathbf{A}_q$ of degree r. Our task in this chapter is to estimate the integral in (6.3) as closely as possible with the goal of establishing that $N(M) > 0$ for sufficiently large values of q^r, $r \geq 5$. In addition to the functions $F_r(t)$ and $H_r(t)$ on $\mathbb{A}_k$ defined in §6.1, we will also need

$$H_{a,b}(t) = \sum_{a \leq \deg P < b}' E(Pt) \tag{7.1}$$

where a and b are positive real numbers with $a < b$. Throughout this chapter, we will use the abbreviation

$$D = D^{r/2}$$

for the fundamental domain defined in (6.25) to (6.27) with $\epsilon = r/2$.

7.1 Preliminary results

We begin our analysis of (6.3) with six preliminary results which will set the stage for the approximations in the next section. Our first lemma generalizes the righthand inequality of (1.23). For any real number x, put

$$L_q(x) \stackrel{\text{def}}{=} \sum_{1 \leq i \leq x} \frac{q^i}{i}.$$

Lemma 7.1. *If $x \geq 1$, then*

$$L_q(x) \leq \frac{3}{2}\left(\frac{q}{q-1}\right)\frac{q^x}{x}. \tag{7.2}$$

Proof. If $1 \le x < 2$, $L_q(x) = q$ and the inequality (7.2) obviously holds for all $q \ge 2$. If $x \ge 2$, then q^x/x has a positive derivative since $q \ge 2$. It is therefore an increasing function of x. Now, the left hand side of (7.2) is constant on the intervals $r \le x < r+1$ for all integers $r \ge 2$. It suffices therefore to prove (7.2) just for integer values of x. This goes by induction as in Lemma 1.24. ■

We must also generalize this same half of (1.23) in a different direction. Releasing 'q' for the moment from its role as a prime power, we define

$$L_{q,a}(r) \stackrel{\text{def}}{=} \sum_{a \le i \le r} \frac{q^i}{i}$$

for all real values $q > 1$ and $a > 0$ and integers $r \ge a$. Then we have:

Lemma 7.2. *If $a > 1/(q-1)$, then*

$$L_{q,a}(r) \le \left(\frac{q}{q-1-a^{-1}}\right)\frac{q^r}{r}.$$

Proof. We proceed by induction on r. If $r = \{a\}$, the least integer greater than or equal to a, then the inequality (7.2) is just

$$\frac{q^r}{r} \le \left(\frac{q}{q-1-a^{-1}}\right)\frac{q^r}{r}$$

which is obviously valid. Now for $r \ge \{a\}$, we must show that

$$\left(\frac{q}{q-1-a^{-1}}\right)\frac{q^r}{r} + \frac{q^{r+1}}{r+1} < \left(\frac{q}{q-1-a^{-1}}\right)\frac{q^{r+1}}{r+1}.$$

Since the left hand side of this inequality equals

$$\left(\frac{q}{q-1-a^{-1}}\right)\frac{q^{r+1}}{r+1}\left(1 + \frac{1-r/a}{rq}\right),$$

and since $r \ge a$, the proof is complete. ■

Lemma 7.3. *Suppose that $t \in \mathbb{A}_k$ and $\chi \in \widehat{W}$ have the property that $|G_v^*(\chi_v, t_v)| \ne 0$ for all $v \in \mathbf{M}$. Then*

1. *if $m(\chi_\infty) \ne 1$, then $\mathfrak{m}(\chi)|\mathfrak{n}(t)$; and*
2. *if $m(\chi_\infty) = 1$, then $\mathfrak{m}(\chi)|\mathfrak{n}(t)\infty$.*

Proof. Let v be a finite place. If $m(\chi_v) \geq 1$, then by Theorem 3.4 $|G_v(\chi_v, t_v)| \neq 0$ implies that $\mathfrak{m}_v(\chi_v) = \mathfrak{n}_v(t_v)$. Hence $\tilde{\mathfrak{m}}(\chi)|\tilde{\mathfrak{n}}(t)$. Now if $m(\chi_\infty) > 1$, then by Theorem 3.7 $\mathfrak{m}_\infty(\chi_\infty) = \mathfrak{n}_\infty(t_\infty)$ (since $m(\chi_\infty) = n(E^t_\infty) = 2 - v_\infty(t_\infty)$). If $m(\chi_\infty) \leq 1$, then $|G^{(1)}_\infty(\chi_\infty, t_\infty)| \neq 0 \Rightarrow 2 - v_\infty(t_\infty) \leq 1 \Rightarrow v_\infty(t_\infty) \geq 1 \Rightarrow \mathfrak{n}_\infty(t_\infty) = 1$ by Theorem 3.5. Hence $m(\chi_\infty) \neq 1 \Rightarrow \mathfrak{m}(\chi)|\mathfrak{n}(t)$ and $m(\chi_\infty) = 1 \Rightarrow \mathfrak{m}(\chi)|\mathfrak{n}(t)\infty$. ∎

Lemma 7.4. *Let χ be a non-trivial character of W (equivalently, of J_k of Dirichlet type) which satisfies one of the following two conditions:* (**i**) $m(\chi_\infty) = 0$ *and* $\deg \mathfrak{m}(\chi) \leq r/2$, *or* (**ii**) $m(\chi_\infty) \geq 1$ *and* $\deg \mathfrak{m}(\chi) \leq (r+3)/2$. *Then*

1. $|\pi_r(\chi)| \leq \left(\frac{1}{2} + \frac{C_q}{r}\right) q^{r/2}$, *and*
2. $\nu \geq \frac{3}{4}(r-1) \Rightarrow |\pi_\nu(\chi)| \leq \left(\frac{2}{3} + \frac{C_q}{\nu}\right) q^{\nu/2}$

where C_q is as defined in Theorem 5.7.

Proof. By Theorem 5.7, if condition (**i**) holds, then

$$|\pi_r(\chi)| \leq \left(\frac{r}{2} + C_q - 1\right) \frac{q^{r/2}}{r} \leq \left(\frac{1}{2} + \frac{C_q}{r}\right) q^{r/2};$$

and if condition (**ii**) holds, then

$$|\pi_r(\chi)| \leq \left(\frac{r+3}{2} + C_q - 2\right) \frac{q^{r/2}}{r} \leq \left(\frac{1}{2} + \frac{C_q}{r}\right) q^{r/2}.$$

This proves the first inequality. Under condition (**i**) and assuming $\nu \geq \frac{3}{4}(r-1)$, so that $r \leq \frac{4\nu}{3} + 1$, we obtain

$$|\pi_\nu(\chi)| \leq \left[\frac{1}{2}\left(\frac{4\nu}{3} + 1\right) + C_q - 1\right] \frac{q^{\nu/2}}{\nu} \leq \left(\frac{2}{3} + \frac{C_q}{\nu}\right) q^{\nu/2}$$

by Theorem 5.7 again. Under condition (**ii**), we obtain

$$|\pi_\nu(\chi)| \leq \left[\frac{1}{2}\left(\frac{4\nu}{3} + 4\right) + C_q - 2\right] \frac{q^{\nu/2}}{\nu} \leq \left(\frac{2}{3} + \frac{C_q}{\nu}\right) q^{\nu/2}.$$

This proves the second inequality. ∎

We may combine these two lemmas and some results from Chapter 6 to prove the following.

Proposition 7.5. *If $t \in D$ and if $\frac{3}{4}(r-1) \le \nu \le r$, then*

$$\left|F_r(t) - \pi_r(\chi_0)G(\chi_0, t^{(r)})\right| \le \left(\frac{1}{2} + \frac{C_q}{r}\right) q^{r/2} \left[\frac{\Phi\left(\mathfrak{n}(t^{(r)})\right)}{q^{\delta(T^r t_\infty)}}\right]^{\frac{1}{2}}$$

and

$$\left|F_\nu(t) - \pi_\nu(\chi_0)G(\chi_0, t^{(\nu)})\right| \le \left(\frac{2}{3} + \frac{C_q}{\nu}\right) q^{\nu/2} \cdot \Phi\left(\mathfrak{n}(t^{(\nu)})\right)^{\frac{1}{2}}.$$

Proof. We start with the representation (6.9) for $F_r(t)$ and place the χ_0-term on the lefthand side. We need to find a bound for each of the factors $\pi_r(\chi)$, $\chi \ne \chi_0$, for which the corresponding product of Gauss sums is nonzero. By the characterizations of D^{odd} and D^{even} in (6.29) and (6.31), we have

$$\deg \mathfrak{n}(t^{(r)}) \le \begin{cases} (r-1)/2 + 2\delta(T^r t_\infty) & r \text{ odd} \\ r/2 + \delta(T^r t_\infty) & r \text{ even.} \end{cases}$$

If $m(\chi_\infty) \le 1$, then by Theorem 3.5,

$$\begin{aligned} \left|G^{(1)}_\infty(\chi_\infty, T^r t_\infty)\right| \ne 0 \quad &\Rightarrow \quad 2 - \mathrm{v}_\infty(T^r t_\infty) \le 1 \\ &\Rightarrow \quad \mathrm{v}_\infty(T^r t_\infty) \ge 1 \\ &\Rightarrow \quad \delta(T^r t_\infty) = 0 \\ &\Rightarrow \quad \deg \mathfrak{n}(t^{(r)}) \le r/2. \end{aligned}$$

If $m(\chi_\infty) = 0$, then by Lemma 7.3, $\deg \mathfrak{m}(\chi) \le \deg \mathfrak{n}(t^{(r)}) \le r/2$, and so condition (**i**) of Lemma 7.4 is satisfied. If $m(\chi_\infty) = 1$, then by Lemma 7.3

$$\deg \mathfrak{m}(\chi) \le \deg\left(\mathfrak{n}(t^{(r)})\infty\right) \le r/2 + 1 < (r+3)/2,$$

and condition (**ii**) of Lemma 7.4 is satisfied. Finally if $m(\chi_\infty) > 1$, then

$$\deg \mathfrak{m}(\chi) \le \deg \mathfrak{n}(t^{(r)}) \le \max\left\{\frac{r-1}{2} + 2, \frac{r}{2} + 1\right\} = \frac{r+3}{2},$$

and again condition (**ii**) is satisfied. We also observe that

$$\nu \le r \Rightarrow \deg \mathfrak{n}(t^{(\nu)}) \le \deg \mathfrak{n}(t^{(r)}).$$

Thus Lemma 7.4 and Corollary 6.4 apply to give us the desired result. ∎

Proposition 7.5, in contrast with Theorem 6.9, gives a specific value for the error introduced when $F_r(t)$ is replaced by $\pi(\chi_0)G(\chi_0, t)$. Whereas

Theorem 6.9 was used to guide us in the selection of a best possible fundamental domain, it is Proposition 7.5 which will actually be used in deriving an asymptotic formula for $N(M)$.

We remark also that it is because of Proposition 7.5 that we replace the sum $H_r(t)$ by $H_{a,r}(t)$ in what follows. We now look at these sums more closely. Let

$$a = \frac{3}{4}(r-1)$$

and let

$$H^*_{a,r}(t) = \sum_{a \le \nu < r} G(\chi_0, t^{(\nu)})\pi_\nu(\chi_0)\,. \tag{7.3}$$

Proposition 7.6. *If $H_{a,r}(t)$ and $H^*_{a,r}(t)$ are as in (7.1) and (7.3), and if $r \ge 5$, then*

$$\left|H_{a,r}(t) - H^*_{a,r}(t)\right| \le \left(\frac{2}{3} + \frac{B_{q,a}}{r-1}\right)\frac{q^{r/2}}{\sqrt{q}-1}\cdot \Phi\left(\mathfrak{n}(t^{(r)})\right)^{\frac{1}{2}}$$

where

$$B_{q,a} = C_q\left(\frac{\sqrt{q}-1}{\sqrt{q}-1-a^{-1}}\right)\,. \tag{7.4}$$

Proof. Because $r \ge 5$, $a \ge 3$. Now by (7.1), $H_{a,r}(t) = \sum_{a \le \nu < r} F_\nu(t)$. We see therefore by Proposition 7.5 that for $t \in D$,

$$\begin{aligned}
\left|H_{a,r}(t) - H^*_{a,r}(t)\right| &\le \sum_{a \le \nu < r}\left|F_\nu(t) - G(\chi_0, t^{(\nu)})\pi_\nu(\chi_0)\right| \\
&\le \sum_{a \le \nu < r}\left(\frac{2}{3} + \frac{C_q}{\nu}\right) q^{\nu/2}\cdot \Phi\left(\mathfrak{n}(t^{(\nu)})\right)^{\frac{1}{2}} \\
&\le \Phi\left(\mathfrak{n}(t^{(r)})\right)^{\frac{1}{2}}\left(\frac{2}{3}\sum_{3 \le \nu < r} q^{\nu/2} + C_q \sum_{a \le \nu < r}\frac{q^{\nu/2}}{\nu}\right) \\
&\le \left(\frac{2}{3} + \frac{B_{q,a}}{r-1}\right)\frac{q^{r/2}}{\sqrt{q}-1}\cdot \Phi\left(\mathfrak{n}(t^{(r)})\right)^{\frac{1}{2}}
\end{aligned}$$

by Lemma 7.2 with $\sqrt{q}$ replacing q, because $a \ge 3 > 1+\sqrt{2} = 1/(\sqrt{2}-1) \ge 1/(\sqrt{q}-1)$. ∎

We observe from Theorem 5.7 and (7.4) that $B_{q,a}$ with $a = \frac{3}{4}(r-1)$ is a decreasing function of both q and r.

7.2 An asymptotic formula for $N(M)$

In this section we put together many pieces to establish an asymptotic formula for $N(M)$ from the integral representation (6.3). We assume throughout that $r \geq 5$. Since the process is rather lengthy, it may be helpful to outline the steps first. They are as follows.

Step 1: We replace the full sum $H_r(t)$ with the partial sum $H_{a,r}(t)$ where $a = \frac{3}{4}(r-1)$. This produces an error of order $q^{7r/4}$ but is necessary so that the error in Step 2 is tolerable.
Step 2: We replace $F_r(t)$ by $\pi_r(\chi_0)G(\chi_0, t)$ and $H_{a,r}(t)$ by $H^*_{a,r}(t)$ using Propositions 7.5 and 7.6. This also produces an error of order $q^{7r/4}$.
Step 3: We approximate the integral

$$I(M) = \int_D \pi_r(\chi_0)G(\chi_0, t^{(r)})H^*_{a,r}(t)^2 E(-Mt)\,dt\,, \tag{7.5}$$

obtaining (7.27) below which involves the global Radon–Nikodym derivative developed in §4.4. Again, an error of order $q^{7r/4}$ occurs.
Step 4: We replace $\pi_r(\chi_0)$ by q^r/r and $\sum_{a\leq\nu<r}\pi_\nu(\chi_0)$ by $L_q(r-1)$, producing one final error of order $q^{7r/4}$.
Step 5: We put the pieces together, obtaining the asymptotic formula of Theorem 7.10.

7.2.1 Step 1: replacing $H_r(t)$ by $H_{a,r}(t)$

Let $a = \frac{3}{4}(r-1)$, and put

$$\begin{aligned} I_1 &= \int_D F_r(t)H_a(t)H_r(t)E(-Mt)\,dt\,, \\ I_2 &= \int_D F_r(t)H_a(t)^2 E(-Mt)\,dt\,, \\ I_3 &= \int_D F_r(t)H_{a,r}(t)^2 E(-Mt)\,dt\,. \end{aligned}$$

Since $H_{a,r}(t) = H_r(t) - H_a(t)$, (6.3) implies

$$\begin{aligned} N(M) - I_3 &= \int_D F_r(t)\left(H_r(t)^2 - H_{a,r}(t)^2\right)E(-Mt)\,dt \\ &= \int_D F_r(t)H_a(t)\left(H_r(t) + H_{a,r}(t)\right)E(-Mt)\,dt \end{aligned}$$

$$= \quad 2I_1 - I_2\,. \tag{7.6}$$

We wish to obtain upper bounds on the absolute values of I_1 and I_2. Note that $|H_a(t)| \le \sum_{1\le\nu<a}\pi_\nu(\chi_0) \le L_q(a-1/4)$ by (5.11) and since $\nu < a \Rightarrow \nu \le a - 1/4$ by definition of a. Hence by Lemma 7.1, we have

$$|H_a(t)| \le \frac{3}{2}\left(\frac{q}{q-1}\right)\frac{q^{a-1/4}}{a-1/4}.$$

Using this bound in I_1 and applying the Cauchy inequality, we obtain

$$|I_1| \le \frac{3}{2}\left(\frac{q}{q-1}\right)\left(\frac{q^{a-1/4}}{a-1/4}\right)\left[\int_D |F_r(t)|^2\,dt\right]^{\frac{1}{2}}\left[\int_D |H_r(t)|^2\,dt\right]^{\frac{1}{2}}.$$

Now

$$\begin{aligned}\int_D |F_r(t)|^2\,dt &= \int_D F_r(t)\cdot\overline{F_r(t)}\,dt \\ &= \int_D \sum_{\deg P_1,P_2=r}{}' \quad E((P_1-P_2)t)\,dt \\ &= \sum_{P_1=P_2}{}' \; 1 = \pi_r(\chi_0) \le q^r/r\,.\end{aligned}$$

A similar calculation together with Lemma 7.1 leads to the bound

$$\int_D |H_r(t)|^2\,dt \le \frac{3}{2}\left(\frac{q}{q-1}\right)\cdot\frac{q^{r-1}}{r-1}.$$

We have then

$$|I_1| \le \frac{3}{2}\left(\frac{q}{q-1}\right)\left(\frac{q^{3r/4-1}}{3r/4-1}\right)\frac{q^{r/2}}{r^{1/2}}\left[\frac{3}{2}\left(\frac{q}{q-1}\right)\frac{q^{r-1}}{r-1}\right]^{\frac{1}{2}},$$

so that

$$|I_1| \le \frac{3\sqrt{3}}{2\sqrt{2}}\cdot\frac{q^{7r/4}}{(q-1)^{3/2}}\cdot\frac{1}{(3r/4-1)\sqrt{r(r-1)}}. \tag{7.7}$$

In the same way, after applying the trivial bound $|F_r(t)| \le q^r/r$, we get

$$\begin{aligned}|I_2| &\le \int_D |F_r(t)||H_a(t)|^2\,dt \\ &\le \frac{q^r}{r}\cdot L_q(a-1/4)\end{aligned}$$

$$\le \frac{q^r}{r}\cdot\frac{3}{2}\left(\frac{q}{q-1}\right)\left(\frac{q^{3r/4-1}}{3r/4-1}\right),$$

so that

$$|I_2| \le \frac{3q^{7r/4}}{2(q-1)}\cdot\frac{1}{r(3r/4-1)}. \tag{7.8}$$

By (7.6), (7.7), and (7.8), we have then

$$\begin{aligned}|N(M)-I_3| &= |2I_1-I_2| \le 2|I_1|+|I_2| \\ &\le \frac{q^{7r/4}}{(q-1)(r-1)}\left(\frac{3\sqrt{3}(r-1)}{\sqrt{2(q-1)}\,(3r/4-1)\sqrt{r(r-1)}}+\frac{3(r-1)}{2r(3r/4-1)}\right) \\ &= \frac{q^{7r/4}}{(q-1)(r-1)}\left(\frac{6\sqrt{6}\sqrt{r-1}}{\sqrt{q-1}\,(3r-4)\sqrt{r}}+\frac{6(r-1)}{r(3r-4)}\right).\end{aligned}$$

Hence

$$|N(M)-I_3| \le \frac{q^{7r/4}}{(q-1)(r-1)}\cdot C_1(q,r) \tag{7.9}$$

where $C_1(q,r)$ is the quantity in the large parentheses in the last equation above. The reason for writing this bound in this particular form will be apparent after Steps 2, 3, and 4 also produce errors in this form. We observe that $C_1(q,r)$ is a decreasing function of both q and r.

7.2.2 Step 2: replacing $F_r(t)$ and $H_{a,r}(t)$

We use Propositions 7.5 and 7.6 to replace $F_r(t)$ and $H_{a,r}(t)$ by their respective major terms. Since

$$\begin{aligned}F_r(t)H_{a,r}(t)^2 - \pi_r(\chi_0)G(\chi_0,t^{(r)})H^*_{a,r}(t)^2 \\ &= \left[F_r(t)-\pi_r(\chi_0)G(\chi_0,t^{(r)})\right]H_{a,r}(t)^2 \\ &\quad + \left[H_{a,r}(t)-H^*_{a,r}(t)\right]H_{a,r}(t)\pi_r(\chi_0)G(\chi_0,t^{(r)}) \\ &\quad + \left[H_{a,r}(t)-H^*_{a,r}(t)\right]H^*_{a,r}(t)\pi_r(\chi_0)G(\chi_0,t^{(r)}),\end{aligned}$$

and recalling (7.5), we have

$$\begin{aligned}|I_3-I(M)| &\le \int_D \left|F_r(t)H_{a,r}(t)^2-\pi_r(\chi_0)G(\chi_0,t^{(r)})H^*_{a,r}(t)^2\right|\,dt \\ &\le \int_D \left|F_r(t)-\pi_r(\chi_0)G(\chi_0,t^{(r)})\right|\cdot|H_{a,r}(t)|^2\,dt \\ &\quad + \int_D \left|H_{a,r}(t)-H^*_{a,r}(t)\right|\left|H_{a,r}(t)\right|\left|\pi_r(\chi_0)\right|\left|G(\chi_0,t^{(r)})\right|\,dt\end{aligned}$$

$$+\int_D \left|H_{a,r}(t) - H^*_{a,r}(t)\right| \left|H^*_{a,r}(t)\right| |\pi_r(\chi_0)| \left|G(\chi_0, t^{(r)})\right| dt\,.$$

Let

$$I_4 = \int_D \left[\Phi\left(\mathfrak{n}(t^{(r)})\right) \cdot q^{-\delta(T^r t_\infty)}\right]^{\frac{1}{2}} |H_{a,r}(t)|^2\, dt\,,$$

$$I_5 = \int_D \Phi\left(\mathfrak{n}(t^{(r)})\right)^{\frac{1}{2}} \left|H_{a,r}(t) G(\chi_0, t^{(r)})\right| dt\,,$$

and

$$I_6 = \int_D \Phi\left(\mathfrak{n}(t^{(r)})\right)^{\frac{1}{2}} \left|H^*_{a,r}(t) G(\chi_0, t^{(r)})\right| dt\,.$$

By Propositions 7.5 and 7.6 and the bound $\pi_r(\chi_0) \le q^r/r$, we get

$$\begin{aligned} |I_3 - I(M)| \le & \left(\frac{1}{2} + \frac{C_q}{r}\right) \cdot q^{r/2} I_4 \\ & + \left(\frac{2}{3} + \frac{B_{q,a}}{r-1}\right) \frac{q^{r/2}}{\sqrt{q}-1} \cdot \frac{q^r}{r}(I_5 + I_6)\,. \end{aligned} \tag{7.10}$$

We must now obtain upper bounds for the integrals I_4, I_5, and I_6. First consider I_4. From definitions (6.12) and (6.15) and by (6.21), we have

$$\begin{aligned} \Phi\left(\mathfrak{n}(t^{(r)})\right) & \le |\tilde{\mathfrak{n}}(t)| \cdot q^{-\delta(T^r t_\infty)} \left|\mathfrak{n}_\infty\left(T^r t_\infty\right)\right| \\ & \le \left|\mathfrak{n}(t^{(r)})\right| q^{-\delta(T^r t_\infty)} = q^{\deg \mathfrak{n}(t^{(r)}) - \delta(T^r t_\infty)}\,. \end{aligned}$$

Therefore, the characterizations (6.29) and (6.31) of D^{odd} and D^{even} imply that

$$\Phi\left(\mathfrak{n}(t^{(r)})\right) \cdot q^{-\delta(T^r t_\infty)} \le q^{r/2}.$$

We have then (cf. Step 1)

$$\begin{aligned} I_4 & \le q^{r/4} \int_D |H_{a,r}(t)|^2\, dt \le q^{r/4} \int_D |H_r(t)|^2\, dt \\ & \le q^{r/4} \cdot \frac{3}{2}\left(\frac{q}{q-1}\right) \frac{q^{r-1}}{r-1}\,, \end{aligned}$$

and so

$$I_4 \le \frac{3q^{5r/4}}{2(q-1)(r-1)}\,. \tag{7.11}$$

For I_5, we first apply the Cauchy inequality:

$$I_5 \le \left[\int_D \Phi\left(\mathfrak{n}(t^{(r)})\right) \cdot \left|G(\chi_0, t^{(r)})\right|^2 dt\right]^{\frac{1}{2}} \left[\int_D |H_{a,r}(t)|^2\, dt\right]^{\frac{1}{2}}.$$

Therefore

$$I_5 \le \left(\frac{3q^r}{2(q-1)(r-1)}\right)^{\frac{1}{2}} \cdot \sqrt{I_7} \tag{7.12}$$

where

$$I_7 = \int_D \Phi\big(\mathfrak{n}(t^{(r)})\big)\,\big|G\big(\chi_0, t^{(r)}\big)\big|^2\,dt\,. \tag{7.13}$$

Now by Corollary 6.10, $G(\chi_0, t^{(r)}) \neq 0$ only if $\mathrm{v}_\infty(t_\infty) > r$, in which case $\mathfrak{n}(t^{(r)}) = \tilde{\mathfrak{n}}(t) = Q$, so

$$\Phi\big(\mathfrak{n}(t^{(r)})\big) \cdot \big|G\big(\chi_0, t^{(r)}\big)\big|^2 = \begin{cases} \Phi(Q)\frac{\mu^2(Q)}{\Phi^2(Q)} & \text{if } \mathrm{v}_\infty(t_\infty) > r \\ 0 & \text{if } \mathrm{v}_\infty(t_\infty) \le r. \end{cases}$$

But by the definition of $D_Q^{r/2}$, the above quantity will be non-zero precisely on that set of $t \in D_Q^{r/2}$ for which $\mathrm{v}_\infty(t_\infty) \ge r+1$ (since $r+1 \ge j + r/2 + 1$ where $j \le r/2$ is the degree of Q). Hence

$$\begin{aligned} I_7 &= \sum_{0\le j\le r/2} \sideset{}{'}\sum_{\deg Q = j} \int_{D_Q^{r/2}} \Phi\big(\mathfrak{n}(t^{(r)})\big)\,\big|G\big(\chi_0, t^{(r)}\big)\big|^2\,dt \\ &= \sum_{0\le j\le r/2} \sideset{}{'}\sum_{\deg Q = j} \frac{\mu^2(Q)}{\Phi(Q)} \cdot meas\left(\prod_{\mathfrak{p}_\mathrm{v} \nmid Q} \mathcal{O}_\mathrm{v} \times T_Q \times \pi_\infty^{r+1}\mathcal{O}_\infty\right) \\ &= \sum_{0\le j\le r/2} \sideset{}{'}\sum_{\deg Q = j} \frac{\mu^2(Q)}{\Phi(Q)} \cdot \Phi(Q)\, q^{-r}, \end{aligned}$$

and so

$$I_7 = q^{-r} \sum_{0\le j\le r/2} \sideset{}{'}\sum_{\deg Q = j} \mu^2(Q)\,. \tag{7.14}$$

At this point we need to evaluate the number of positive square-free polynomials of degree j.

Lemma 7.7. *The number of square-free positive polynomials in* $\mathbf{A}_q$ *of degree* j *is*

$$\begin{cases} 1 & \text{if } j = 0 \\ q & \text{if } j = 1 \\ q^j - q^{j-1} & \text{if } j \ge 2\,. \end{cases}$$

Proof. See Exercise 7.2(1). ∎

Combining this result with (7.14), we get

$$I_7 = q^{-r}\left[1+q+\sum_{2\le j\le r/2}(q^j-q^{j-1})\right] = q^{-r}(1+q^{[r/2]}),$$

so that

$$I_7 \le q^{-r}(1+q^{r/2}) = q^{-r/2}(q^{-r/2}+1). \tag{7.15}$$

Combining (7.12) and this last inequality, we arrive at the estimate

$$I_5 \le \frac{\sqrt{6}q^{r/4}}{2\sqrt{q-1}\sqrt{r-1}}\sqrt{q^{-r/2}+1}. \tag{7.16}$$

Finally we turn to I_6. We have

$$H^*_{a,r}(t)G(\chi_0,t^{(r)}) = \sum_{a\le\nu<r}\pi_\nu(\chi_0)G(\chi_0,t^{(\nu)})G(\chi_0,t^{(r)}).$$

By Corollary 6.10 again

$$G(\chi_0,t^{(\nu)}) = \frac{\mu(\tilde{\mathfrak{n}}(t))^2}{\Phi(\tilde{\mathfrak{n}}(t))}\cdot E_\infty(T^\nu t_\infty)\cdot X_\nu(t_\infty)$$

where X_ν is the characteristic function of $\pi_\infty^{\nu+1}\mathcal{O}_\infty$. If $Q=\tilde{\mathfrak{n}}(t)$, then

$$G(\chi_0,t^{(\nu)})G(\chi_0,t^{(r)}) = \frac{\mu^2(Q)}{\Phi^2(Q)}\cdot E_\infty\left((T^\nu+T^r)t_\infty\right)\cdot X_r(t_\infty)$$

since $r>\nu$. Hence

$$\begin{aligned}\left|H^*_{a,r}(t)G(\chi_0,t^{(r)})\right| &\le \sum_{a\le j<r}\pi_j(\chi_0)\left|G(\chi_0,t^{(\nu)})\right|\left|G(\chi_0,t^{(r)})\right|\\ &\le \left|G(\chi_0,t^{(r)})\right|^2\sum_{a\le j<r}q^j/j\\ &\le \left|G(\chi_0,t^{(r)})\right|^2\cdot L_q(r-1).\end{aligned}$$

Using this inequality, Lemma 7.1, and Cauchy's inequality, we get

$$\begin{aligned}I_6 &\le \frac{3q^r}{2(q-1)(r-1)}\cdot\int_D\Phi\left(\mathfrak{n}(t^{(r)})\right)^{\frac12}\left|G(\chi_0,t^{(r)})\right|^2dt\\ &\le \frac{3q^r}{2(q-1)(r-1)}\cdot\left[\int_D\Phi\left(\mathfrak{n}(t^{(r)})\right)\left|G(\chi_0,t^{(r)})\right|^2dt\right]^{\frac12}\\ &\quad\times\left[\int_D\left|G(\chi_0,t^{(r)})\right|^2dt\right]^{\frac12}.\end{aligned}$$

Now the integral in the first bracket is just I_7 of (7.13).

The second integral is

$$\begin{aligned}\int_D \left|G(\chi_0, t^{(r)})\right|^2 dt &= \sum_{\deg Q \le r/2}{}' \frac{\mu^2(Q)}{\Phi^2(Q)} \cdot \Phi(Q)\, q^{-r} \\ &= q^{-r} \sum_{\deg Q \le r/2}{}' \frac{\mu^2(Q)}{\Phi(Q)}. \qquad (7.17)\end{aligned}$$

We may now combine Lemma 4.17 with (7.17) to obtain

$$\begin{aligned}\int_D \left|G(\chi_0, t^{(r)})\right|^2 dt &= q^{-r} \sum_{0 \le j \le r/2} \sum_{\deg Q = j}{}' \frac{1}{\Phi(Q)} \\ &\le q^{-r} \sum_{0 \le j \le r/2} j+1 \\ &\le q^{-r} \frac{(r/2+1)(r/2+2)}{2} \\ &= q^{-r} \cdot \frac{(r+2)(r+4)}{8}.\end{aligned}$$

Using the estimate (7.15) for I_7, we get finally

$$\begin{aligned}I_6 &\le \frac{3q^r}{2(q-1)(r-1)} \sqrt{q^{-r/2}(1+q^{-r/2})} \cdot q^{-r/2} \sqrt{(r+2)(r+4)/8} \\ &\le \frac{3q^{r/4}}{2(q-1)(r-1)} \cdot \sqrt{q^{-r/2}+1} \cdot \sqrt{(r+2)(r+4)/8}.\end{aligned}$$

Step 2 of our process can now be completed by putting together the information of (7.10), (7.11), (7.16), and this bound for I_6. We find that

$$\begin{aligned}|I_3 - I(M)| \le{}& \left(\frac{1}{2} + \frac{C_q}{r}\right) q^{r/2} \cdot \frac{3q^{5r/4}}{2(q-1)(r-1)} \\ &+ \left(\frac{2}{3} + \frac{B_{q,a}}{r-1}\right) \times \frac{q^{r/2}}{\sqrt{q}-1} \frac{q^r}{r} \left[\frac{\sqrt{6}q^{r/4}}{2\sqrt{q-1}\sqrt{r-1}} \cdot \sqrt{q^{-r/2}+1}\right] \\ &+ \left(\frac{2}{3} + \frac{B_{q,a}}{r-1}\right) \times \frac{q^{r/2}}{\sqrt{q}-1} \frac{q^r}{r} \left[\frac{3q^{r/4}\sqrt{q^{-r/2}+1}\sqrt{(r+2)(r+4)/8}}{2(q-1)(r-1)}\right] \\ ={}& \frac{q^{7r/4}}{(q-1)(r-1)} \cdot C_2(q,r)\end{aligned}$$

where

$$C_2(q,r) = \frac{3}{2}\left(\frac{1}{2} + \frac{C_q}{r}\right)$$

$$+\left(\frac{2}{3}+\frac{B_{q,a}}{r-1}\right)\cdot\frac{\sqrt{6}\sqrt{q-1}\sqrt{r-1}\sqrt{q^{-r/2}+1}}{2r(\sqrt{q}-1)}$$
$$+\left(\frac{2}{3}+\frac{B_{q,a}}{r-1}\right)\cdot\frac{3\sqrt{2}\sqrt{q^{-r/2}+1}\sqrt{(r+2)(r+4)}}{8r(\sqrt{q}-1)}.$$

Thus we have shown that

$$|I_3 - I(M)| \leq \frac{q^{7r/4}}{(q-1)(r-1)}\cdot C_2(q,r) \tag{7.18}$$

where $C_2(q,r)$ is defined immediately above, C_q is as in Theorem 5.7, and $B_{q,a}$ is given by (7.4). Like $C_1(q,r)$, $C_2(q,r)$ is a decreasing function of both q and r. We have completed Step 2 of our process.

7.2.3 Step 3: estimating the integral of major terms

By the inequalities in (7.9) and (7.18), we see that $N(M)$ and $I(M)$ differ by an error term of order $q^{7r/4}$. We wish now to analyse further $I(M)$. From the definition (7.5), we have

$$I(M) = \pi_r(\chi_0)\sum_{a\leq\nu_1,\nu_2<r}\pi_{\nu_1}(\chi_0)\pi_{\nu_2}(\chi_0)\cdot I_8(\nu_1,\nu_2) \tag{7.19}$$

where

$$I_8(\nu_1,\nu_2) = \int_D G(\chi_0,t^{(r)})G(\chi_0,t^{(\nu_1)})G(\chi_0,t^{(\nu_2)})E(-Mt)\,dt.$$

Let $\widetilde{G}(\chi_0,\tilde{t})$ be the product over $\mathrm{v}\in\mathbf{M}_{\mathrm{fin}}$ in (6.3). Splitting the components D_Q of the fundamental domain D into their finite and infinite parts (see (6.26) and (6.27)), we may write

$$\begin{aligned} I_8(\nu_1,\nu_2) &= \sideset{}{'}\sum_{\deg Q\leq r/2}\int_{\widetilde{D}_Q}\widetilde{G}(\chi_0,\tilde{t})^3E(-M\tilde{t})\,d\tilde{t} \\ &\times\int_{\pi_\infty^{r+1}\mathcal{O}_\infty}E_\infty\left((T^r+T^{\nu_1}+T^{\nu_2}-M)t_\infty\right)dt_\infty \end{aligned} \tag{7.20}$$

because $X_{\nu_1}X_{\nu_2}X_r = X_r$ and the infinity component of any $D_Q\subseteq D$ equals

$$\pi_\infty^{[r/2]+\deg Q+1}\mathcal{O}_\infty \supseteq \pi_\infty^{r+1}\mathcal{O}_\infty$$

as $\deg Q\leq r/2$. Now $\deg M = r \Rightarrow \deg(T^r+T^{\nu_1}+T^{\nu_2}-M)\leq r-1$, and so $\mathrm{v}_\infty((T^r+T^{\nu_1}+T^{\nu_2}-M)t_\infty)\geq(1-r)+(r+1)=2$. But $n(E_\infty)=2$, so the above integral is just $meas(\pi_\infty^{r+1}\mathcal{O}_\infty) = q^{-r}$. We observe that this is the only place in the entire process where we use the hypothesis that M is positive of degree r.

Combining this argument with (7.20), we conclude that

$$I_8 = I_8(\nu_1, \nu_2) = q^{-r} \int_{\widetilde{D}} \widetilde{G}(\chi_0, \tilde{t})^3 E(-M\tilde{t})\, d\tilde{t}$$

is independent of ν_1 and ν_2; and this together with (7.19) yields

$$I(M) = q^{-r} \pi_r(\chi_0) \varpi_q(r)^2 \int_{\widetilde{D}} \widetilde{G}(\chi_0, \tilde{t})^3 E(-M\tilde{t})\, d\tilde{t} \tag{7.21}$$

where

$$\varpi_q(r) = \sum_{a \le \nu < r} \pi_\nu(\chi_0)\,. \tag{7.22}$$

We will see that the factor before the integral in (7.21) has the approximate magnitude $L_q(r-1)^2/r$ and that the integral is approximately the global Radon–Nikodym derivative developed in §4.4. We now turn our attention to this integral.

Let

$$V = \prod_{P \in \mathbf{M}_{\text{fin}}} \mathbf{U}_P^3$$

as in §4.4. By definition of the global Gauss sum (6.4), we have

$$\begin{aligned} \widetilde{G}(\chi_0, \tilde{t})^3 &= \int_V E\left((u_1 + u_2 + u_3)\tilde{t}\right) du_1 du_2 du_3 \\ &= \int_{\mathcal{O}} \theta(z; h, V) E(\tilde{t}z)\, dz = \hat{\theta}(\tilde{t}; h, V) \end{aligned}$$

where we have changed variables, giving rise to the global Radon–Nikodym derivative $\theta(z; h, V)$, whose existence for the function

$$h : (u_1, u_2, u_3) \mapsto u_1 + u_2 + u_3$$

was established in §4.4. Throughout the remainder of this chapter, h is the function above. We have shown so far that

$$I(M) = q^{-r} \pi_r(\chi_0) \varpi_q(r)^2 \int_{\widetilde{D}} \hat{\theta}(\tilde{t}; h, V) E(-M\tilde{t})\, d\tilde{t}\,. \tag{7.23}$$

Now by Fourier inversion (Proposition 4.10),

$$\theta(M; h, V) = \int_{\tilde{\mathbf{A}}_k} \hat{\theta}(\tilde{t}; h, V) E(-M\tilde{t})\, d\tilde{t}\,.$$

Since $\tilde{\mathbf{A}}_k$ is the disjoint union of the $\widetilde{D}_Q$, $Q \in \mathbf{A}_q$, Q positive,

$$\theta(M; h, V) = \int_{\widetilde{D}} \hat{\theta}(\tilde{t}; h, V) E(-M\tilde{t})\, d\tilde{t}$$

$$+ \sum_{\deg Q > r/2}' \int_{\widetilde{D}_Q} \hat{\theta}(\tilde{t}; h, V) E(-M\tilde{t})\, d\tilde{t}. \qquad (7.24)$$

Looking at each of the integrals in the summation and applying eqn (4.31), we compute

$$\int_{\widetilde{D}_Q} |\hat{\theta}(\tilde{t}; h, V)|\, d\tilde{t} = \frac{|\mu(Q)|^3}{\Phi(Q)^3} \cdot meas(\widetilde{D}_Q) = \frac{|\mu(Q)|}{\Phi(Q)^2} \qquad (7.25)$$

where the last equality is as in the argument preceding (7.14).

We wish now to bound the difference

$$\left| \theta(M; h, V) - \int_{\widetilde{D}} \hat{\theta}(\tilde{t}; h, V) E(-M\tilde{t})\, d\tilde{t} \right|.$$

In addition, in Step 4 we will need an upper bound on $|\theta(M; h, V)|$. We get both of these bounds in the next result.

Proposition 7.8. *Let $\alpha = 2$ if $q = 2$, and otherwise let $\alpha = 3/4$. Then*

$$|\theta(M; h, V)| \le C_{31} = \frac{\alpha}{(1 - q^{-1/2})^2}, \qquad (7.26)$$

and

$$\left| \theta(M; h, V) - \int_{\widetilde{D}} \hat{\theta}(\tilde{t}; h, V) E(-M\tilde{t})\, d\tilde{t} \right| \le q^{-r/4} C_{32}$$

where

$$C_{32} = \alpha \frac{r(1 - q^{-1/2})/2 + 1}{(1 - q^{-1/2})^2}.$$

Proof. We employ Lemmas 4.16 and 4.17. For the first inequality we have, by (7.24) and (7.25),

$$|\theta(M; h, V)| \le \alpha \sum_{j=0}^{\infty} (j+1) q^{-j/2} = \frac{\alpha}{(1 - q^{-1/2})^2}.$$

For the second inequality, let $w = [r/2]$. Then

$$\left| \theta(M; h, V) - \int_{\widetilde{D}} \hat{\theta}(\tilde{t}; h, V) E(-M\tilde{t})\, d\tilde{t} \right| \le \sum_{\deg Q > r/2}' \frac{1}{\phi^2(Q)}$$

$$\le \alpha \sum_{j=w+1}^{\infty} (j+1) q^{-j/2} = \alpha q^{(w+1)/2} \left[\frac{(w+1)(1 - q^{-1/2}) + 1}{(1 - q^{-1/2})^2} \right].$$

Now one checks easily that this last expression is a decreasing function of both q and w. Replacing w with $r/2 - 1$, we obtain our result. ∎

We are now in a position to complete Step 3. By (7.23) and Proposition 7.8, we have

$$\left|I(M) - q^{-r}\pi_r(\chi_0)\varpi_q(r)^2\theta(M;h,V)\right| \leq q^{-r}\pi_r(\chi_0)\varpi_q(r)^2 \cdot q^{-r/4}C_{32}.$$

Since $\pi_r(\chi_0) \leq q^r/r$ and $\varpi_q(r) \leq L_q(r-1)$, this inequality and Lemma 7.1 imply

$$\left|I(M) - q^{-r}\pi_r(\chi_0)\varpi_q(r)^2\theta(M;h,V)\right| \leq \frac{q^{7r/4}}{(q-1)(r-1)}C_3(q,r) \quad (7.27)$$

where

$$C_3(q,r) = \frac{9C_{32}}{4r(q-1)(r-1)} = \frac{9\alpha\left(r(1-q^{-1/2})/2+1\right)}{4(q-1)r(r-1)(1-q^{-1/2})^2}. \quad (7.28)$$

As the reader will easily verify, $C_3(q,r)$ is a decreasing function of both q and r.

7.2.4 Step 4: the final replacements

We now replace the quantities $\pi_r(\chi_0)$ and $\varpi_q(r)$ in (7.27), arriving finally at a major term of the form $\frac{1}{r}L_q(r-1)^2\theta(M;h,V)$. We need the following.

Proposition 7.9. *Let $\pi_r(\chi_0)$ be the number of positive irreducibles of degree r in $\mathbf{A}_q$. Then*

$$|\pi_r(\chi_0) - q^r/r| < \tfrac{1}{2}q^{r/2},$$

and

$$|\varpi_q(r) - L_q(r-1)| \leq q^{3r/4}C_{41}$$

where

$$C_{41} = \frac{q^{-r/4} - q^{-3(r+1)/8}}{\sqrt{q}-1} + \frac{6}{(q-1)(3r-4)}.$$

Proof. The first part follows from the standard formula

$$\pi_r(\chi_0) = \frac{1}{r}\sum_{d|r}\mu(r/d)\,q^d$$

which follows by Möbius inversion from (5.11). We have

$$|\pi_r(\chi_0) - q^r/r| \leq \frac{1}{r}\sum_{\substack{d|r\\ d\neq r}} q^d \leq \frac{q^{r/2}}{r}\cdot\frac{r}{2} \leq \frac{1}{2}q^{r/2}$$

as required. For the second part we have

$$|\varpi_q(r) - L_q(r-1)| \leq \sum_{a\leq\nu<r}|\pi_\nu(\chi_0) - q^\nu/\nu| + \sum_{1\leq\nu<a} q^\nu/\nu$$

$$\leq \frac{1}{2} \sum_{a\leq\nu<r} q^{\nu/2} + L_q(a-1/4).$$

From the upper bound for $L_q(r)$ in Lemma 7.1,

$$\begin{aligned}\left|\sum_{a\leq\nu<r} \pi_\nu(\chi_0) - L_q(r-1)\right| &\leq \frac{q^{r/2}-q^{a/2}}{\sqrt{q}-1} + \left(\frac{3q}{2(q-1)}\right)\cdot\frac{q^{a-(1/4)}}{a-(1/4)} \\ &= \frac{q^{r/2}-q^{3(r-1)/8}}{\sqrt{q}-1} + \frac{3}{2}\frac{1}{(q-1)}\frac{q^{3r/4}}{(3r/4-1)} \\ &= q^{3r/4}C_{41}\end{aligned}$$

as required. ■

We may now complete Step 4 using Lemma 7.1 and Propositions 7.8 and 7.9. First

$$\begin{aligned}q^{-r}\pi_r(\chi_0)\varpi_q(r)^2 - \frac{1}{r}L_q(r-1)^2 &= \left(q^{-r}\pi_r(\chi_0) - \frac{1}{r}\right)\varpi_q(r)^2 \\ &\quad + \frac{1}{r}\left[\varpi_q(r) - L_q(r-1)\right]\left[\varpi_q(r) + L_q(r-1)\right]\end{aligned}$$

so that

$$\begin{aligned}&\left|q^{-r}\pi_r(\chi_0)\varpi_q(r)^2 - \frac{1}{r}L_q(r-1)^2\right| \\ &\quad\leq \frac{1}{2}q^{-r/2}L_q(r-1)^2 + \frac{1}{r}q^{3r/4}C_{41}\cdot 2L_q(r-1) \\ &\quad= \frac{q^{7r/4}}{(q-1)(r-1)}\left[\frac{9q^{-r/4}}{8(q-1)(r-1)} + \frac{3C_{41}}{r}\right].\end{aligned}$$

We conclude from this inequality together with (7.26) and (7.27) that

$$\left|I(M) - \frac{1}{r}L_q(r-1)^2\theta(M;h,V)\right| \leq \frac{q^{7r/4}}{(q-1)(r-1)}(C_3(q,r)+C_4(q,r))$$

where

$$C_4(q,r) = \left[\frac{9q^{-r/4}}{(q-1)(r-1)} + \frac{3C_{41}}{r}\right]C_{31}.$$

Once again, we observe that $C_4(q,r)$ is a decreasing function of both q and r, an easy check for the reader. This completes Step 4.

7.2.5 Step 5: putting it all together

The hard work is all done, and now we put the pieces together. Let $L_q(r-1)$ and $\theta(M;h,V)$ be as defined in (1.19) and §4.4 respectively. Then combining the estimates (7.9), (7.18), and the last inequality from Step 4, we obtain the following fundamental result:

Theorem 7.10. *We have*

$$\left|N(M) - \frac{1}{r}L_q(r-1)^2\theta(M;h,V)\right| \le \frac{q^{7r/4}}{(q-1)(r-1)}C(q,r)$$

where

$$C(q,r) = C_1(q,r) + C_2(q,r) + C_3(q,r) + C_4(q,r) \tag{7.29}$$

is the sum of the error functions introduced in Steps 1–4 above.

Throughout our derivation of this approximation to $N(M)$, we have tracked the error terms with great care. The payoff for our efforts is Table A.1 of the Appendix which gives tight numerical bounds on q and r that imply $N(M) > 0$.

Exercises 7.2

1. Use the technique of Exercise 5.4(1) to establish Lemma 7.7. Show that

$$Z_{\mu^2}(s) = \prod_P 1 + |P|^{-s} = \prod_P \left(\frac{1-|P|^{-2s}}{1-|P|^{-s}}\right) = \frac{Z_k^*(s)}{Z_k^*(2s)}\,.$$

7.3 The 3-primes asymptotic theorem

As we have observed in Steps 1 to 4, the function $C(q,r)$ defined by (7.29) is a decreasing function of both q and r. We set

$$C = C(2,5) = 36.56\,. \tag{7.30}$$

It follows from Theorem 7.10 that $N(M)$ is positive for $r \ge 5$ provided that

$$\frac{1}{r}L_q(r-1)^2\theta(M;h,V) > \frac{Cq^{7r/4}}{(q-1)(r-1)}\,.$$

If M is *even*, then $\theta(M;h,V) = 0$. Assume therefore that M is odd. We have already established lower bounds for $L_q(r-1)$ (Lemma 1.24) and for

$\theta(M; h, V)$ when M is odd (Proposition 5.11). By those results and the value of C in (7.30), $N(M) > 0$ provided that

$$\frac{0.37}{r}\left(\frac{q^{2r}}{(q-1)^2(r-1)^2}\right) > \frac{36.56q^{7r/4}}{(q-1)(r-1)}.$$

Simplifying, we get

$$N(M) > 0 \text{ provided that } \frac{q^{r/4}}{r(q-1)(r-1)} > 98.81. \qquad (7.31)$$

It is clear that if we fix $r \geq 5$ and then allow $q \to \infty$, the inequality in (7.31) will be satisfied for all sufficiently large q; and the same conclusion holds if we fix q and let $r \to \infty$. In fact, if we replace $q - 1$ by q in the denominator in (7.31), then solving for q we obtain

$$q \geq \left[98.81(r^2 - r)\right]^{\frac{4}{r-4}}. \qquad (7.32)$$

Given $r \geq 5$, we can define q_r to be the least prime power for which (7.32) holds true. One checks easily that q_r is a decreasing function of r and that $q_r = 2$ for all sufficiently large r.

We can at last write down the result which has been the goal of the last two chapters.

Asymptotic Theorem 7.11. *For every degree $r \geq 5$, there exists a q_r such that if $q \geq q_r$, then every odd, positive polynomial $M \in \mathbf{A}_q$ of degree r can be written as*

$$M = P_1 + P_2 + P_3$$

where each P_i is a positive irreducible in $\mathbf{A}_q$ and $\deg P_1 = r$, $\deg P_2 < r$, and $\deg P_3 < r$. Further, the sequence $\{q_r\}_{r \geq 5}$ is decreasing, and $q_r = 2$ for all sufficiently large r.

The reader is referred to the Appendix for a calculation of specific upper bounds on the values of the q_r and a discussion of the work which has led to a complete solution to the polynomial 3-primes problem.

A sea of details has stood between the statement of the 3-primes problem in Chapter 1 and its (partial) solution here. We should note that these techniques of course apply to a large variety of additive problems in the arithmetic of polynomials over a finite field. For example, we embark in the next chapter upon a solution to the polynomial Waring problem.

Exercises 7.3

1. Show that for every $r \geq 1$, the even polynomial $T^{4r} + T^{2r} \in \mathbf{A}_2$ cannot be written as a sum of three irreducibles. Hence the hypothesis that M be odd is necessary.

2. Verify in (7.32) that q_r is a decreasing function of r and that $q_r = 2$ for all sufficiently large r.

8

The polynomial Waring problem

In this chapter, we apply the adèlic circle method to the polynomial Waring problem stated in §1.1. In order to prove Theorem 1.9, it suffices by Lemma 1.12 to show that there is an absolute constant C such that every $M \in \mathbf{A}_q$ with $|M| > C$ is the strict sum of $s = d^2 2^d$ d-th powers. In §8.3, we will deduce this result from Theorem 8.1 below, which will be the main focus of our efforts. As in §1.1, $\{x\}$ denotes the least integer which is greater than or equal to the real number x.

Throughout this chapter, d is a fixed integer, and $\mathbf{A}_q$ is a polynomial ring with $p = \text{char}(\mathbf{A}_q) > d$. Except in Proposition 8.5 below, we assume that $d \geq 2$. For a given polynomial $M \in \mathbf{A}_q$ and a given integer $s > 0$, let $\{\gamma_i \in \mathbf{A}_q^\times\}_{i=1}^s$ be a sequence of non-zero constants such that

$$\gamma \stackrel{\text{def}}{=} \sum_{i=1}^{s} \gamma_i = \text{coefficient of the } nd\text{-th term of } M \tag{8.1}$$

where $n = \{\deg(M)/d\}$. Further, let $N_{d,s}(M)$ be the number of positive polynomials $X_1, \ldots, X_s \in \mathbf{A}_q$, each of degree n, such that

$$M = \gamma_1 X_1^d + \gamma_2 X_2^d + \cdots + \gamma_s X_s^d .$$

We observe that if $\deg M < nd$, then $\gamma = 0$. Let $\theta(\tilde{t}; h, \mathcal{O}^s)$ be the Radon–Nikodym derivative of the map $h : \mathcal{O}^s \to \mathcal{O}$ defined by

$$h(\tilde{t}_1, \tilde{t}_2, \ldots, \tilde{t}_s) = \gamma_1 \tilde{t}_1^d + \gamma_2 \tilde{t}_2^d + \cdots + \gamma_s \tilde{t}_s^d . \tag{8.2}$$

If $s > 2d$, then (see §4.4.2) this derivative exists as a continuous function on $\mathcal{O}$.

Theorem 8.1. (Asymptotic Waring) *If $s \geq d2^d$, then there exists a $\delta > 0$ such that*

$$N_{d,s}(M) = q^{n(s-d)} \cdot \theta(M; h, \mathcal{O}^s) + O\Big(q^{n(s-d-\delta)}\Big) \tag{8.3}$$

where the constant implied by the O-notation depends upon s, d, and δ but is independent of n and q.

It would be more natural to develop an asymptotic formula for the number of representations of the polynomial M in the form

$$M = X_1^d + X_2^d + \cdots + X_s^d$$

with $X_1, X_2, \ldots, X_s$ arbitrary polynomials of degree less than or equal to $\{\deg M/d\}$. This can be done by our methods, but it would involve considerably more effort and technical details. Since our principal aim here is just Theorem 1.9, we have chosen to concentrate our efforts on (8.3), which is interesting in itself.

The circle method provides a surprisingly smooth and beautiful proof of Theorem 8.1 that makes use of much of the theory developed in previous chapters. Before presenting the details of the proof in §§8.1–8.3, we pause to outline the main steps in our argument.

(a) Using the additive character E on $\mathbb{A}_k$ and the fundamental domain $D^{n(d-1)}$ defined in (6.25)–(6.27), we write down an integral formula for $N_{d,s}(M)$.

(b) We divide the fundamental domain $D^{n(d-1)}$ into 'major arcs' and 'minor arcs' in a very natural way. We show that on the union of major arcs, the finite part of the integral formula yields the factor $\theta(M; h, \mathcal{O}^s)$ in the main term of (8.3) whereas both the finite and the ∞-parts contribute to the $q^{n(s-d)}$ factor.

(c) We show that on the minor arcs, a polynomial version of the classical Weyl inequality holds true, so that the magnitude of the integral on the union of these minor arcs is strictly less than the major term of (8.3) under suitable conditions on s and n.

(d) We apply the results of §§3.4 and 4.4, observing in particular that $\theta(M; h, \mathcal{O}^s)$ is bounded below by a positive absolute constant. Theorem 8.1 and then Theorem 1.9 follow.

8.1 Major arc analysis

Let

$$g(t) = \sideset{}{'}\sum_{\deg A = n} E(A^d t) \tag{8.4}$$

where $t \in \mathbb{A}_k$ and where E is the adèlic character of Definition 4.7. Then if $D \subset \mathbb{A}_k$ is any fundamental domain for $\mathbb{A}_k/k$, we have

$$N_{d,s}(M) = \int_D g(\gamma_1 t) g(\gamma_2 t) \cdots g(\gamma_s t) E(-Mt)\, dt \tag{8.5}$$

(cf. §1.4). The most convenient choice of fundamental domain for the proof of Theorem 8.1 is

$$D = D^{n(d-1)}$$

which is defined in (6.25) through (6.27) with $\epsilon = n(d-1)$. We will make considerable use of the decomposition

$$D = D^{n(d-1)} = \bigcup_{\deg Q \leq n(d-1)} D_Q^{n(d-1)}$$

in what follows.

Our job now is to analyse the integral in (8.5). Following the circle method approach to the classical Waring problem, which involves the Farey dissection of the unit interval, we decompose our fundamental domain D as follows:

Definition 8.2. *If* $\deg Q \leq n$, *then* D_Q *is called a* major arc. *If* $n < \deg Q \leq n(d-1)$, *then* D_Q *is called a* minor arc.

Hence $t \in D$ lies in a major arc if and only if $\tilde{\mathfrak{n}}(t) = Q$ has degree less than or equal to n, where $\tilde{\mathfrak{n}}$ is the finite part of the conductor of t (see (6.11) and (6.14)). Our analysis will follow the usual line of evaluating the integral in (8.5) on the major arcs and then showing that the magnitude of the integral on the minor arcs is strictly less than its magnitude on the major arcs under suitable conditions on s and n.

We now embark upon an analysis of the major arc portion of the integral in (8.5). We begin with the following.

Proposition 8.3. *If t lies in a major arc, so that* $\tilde{\mathfrak{n}}(t) = Q$ *with* $\deg Q = j \leq n$, *then*

$$g(t) = \sum_{\deg C < j} \widetilde{E}(\tilde{t}C^d) \sum_{\deg B = n-j}' E_\infty(t_\infty B^d Q^d)$$

where by $\widetilde{E}$ *we mean the restriction of E to the finite adèles* $\tilde{\mathbb{A}}_k$.

Proof. First, let $\mathrm{v} \in \mathbf{M}_{\mathrm{fin}}$, and let $A = BQ + C$ with $\deg C < \deg Q$. Since $Qt_\mathrm{v} \in \mathcal{O}_\mathrm{v}$, we have $E_\mathrm{v}(Qt_\mathrm{v}) = 1$, and so

$$E_\mathrm{v}(t_\mathrm{v} A^d) = E_\mathrm{v}\left(t_\mathrm{v}(BQ + C)^d\right) = E_\mathrm{v}\left(t_\mathrm{v} C^d\right).$$

Now, let $\mathrm{v} = \infty$. We have,

$$\mathrm{v}_\infty\left(t_\infty((BQ + C)^d - (BQ)^d)\right) = \mathrm{v}_\infty(t_\infty) - \deg\left((BQ + C)^d - (BQ)^d\right)$$

$$\begin{aligned} &\geq \mathrm{v}_\infty(t_\infty) - (n(d-1) + (j-1)) \\ &\geq n(d-1) + j + 1 - (n(d-1) + j - 1) \\ &= 2\,, \end{aligned}$$

where in the last inequality we have used the definition of D_Q. Since $n(E_\infty) = 2$, we see that $E_\infty(t_\infty A^d) = E_\infty(t_\infty B^d Q^d)$. Thus

$$\begin{aligned} g(t) &= \sum_{\deg A=n}{}' E(tA^d) \\ &= \sum_{\deg A=n}{}' \left(\prod_{\mathrm{v}\neq\infty} E_\mathrm{v}(t_\mathrm{v} A^d) \right) E_\infty(t_\infty A^d) \\ &= \sum_{\deg C<j} \widetilde{E}(\tilde{t}C^d) \sum_{\deg B=n-j}{}' E_\infty(t_\infty B^d Q^d) \end{aligned}$$

as claimed. ■

This result says that on D_Q the evaluation of $E(At)$ depends on the infinite evaluation of the 'whole part' of A (upon division by Q) and on the finite evaluation of the remainder. We now look at each of these parts more closely.

Proposition 8.4. *We have*

$$\sum_{\deg C<j} \widetilde{E}(\tilde{t}C^d) = q^j \int_{\mathcal{O}} \widetilde{E}(\tilde{t}z^d)\,dz\,,$$

where $\mathcal{O}$ is the compact ring (4.14).

Proof. The ideal $Q\mathcal{O}$ of $\mathcal{O}$ has measure q^{-j} and $\bigcup_{\deg C<j}(C + Q\mathcal{O}) = \mathcal{O}$. Moreover since $t \in D_Q$, $\widetilde{E}(\tilde{t}z) = 1$ for all $z \in Q\mathcal{O}$, so we may write

$$\sum_{\deg C<j} \widetilde{E}(\tilde{t}C^d) = \sum_{\deg C<j} q^j \int_{C+Q\mathcal{O}} E(\tilde{t}z^d)\,dz = q^j \int_{\mathcal{O}} E(\tilde{t}z^d)\,dz$$

as desired. ■

The infinite part is more challenging.

Proposition 8.5. *For fixed $d \geq 1$, suppose that $y \in k_\infty$ satisfies $\mathrm{v}_\infty(y) > m(d-1)$. Then*

$$\sum_{\deg B=m}{}' E_\infty(yB^d) = \begin{cases} q^m E_\infty(yT^{md}) & \text{if } \mathrm{v}_\infty(y) > md \\ 0 & \text{otherwise.} \end{cases}$$

Proof. Let $w \in \pi_\infty^{m+1}\mathcal{O}_\infty$. Then since $n(E_\infty) = 2$ and

$$\begin{aligned} \mathrm{v}_\infty(yT^{md}w) &= \mathrm{v}_\infty(y) + \mathrm{v}_\infty(T^{md}) + \mathrm{v}_\infty(w) \\ &> m(d-1) - md + m + 1 = 1\,, \end{aligned}$$

$E_\infty(yB^d) = E_\infty(yT^{md}((B/T^m)^d + w))$ for any positive B of degree m. Thus since $meas(\pi_\infty^{m+1}\mathcal{O}_\infty) = q^{-m}$, we get

$$E_\infty(yB^d) = q^m \int_{(B/T^m)^d + \pi_\infty^{m+1}\mathcal{O}_\infty} E_\infty(yT^{md}u)\, du\,.$$

We will now show that the cosets $(B/T^m)^d + \pi_\infty^{m+1}\mathcal{O}_\infty$ are pairwise disjoint. For B_1 and B_2 positive of degree m, $B_1 \neq B_2$, we have

$$\begin{aligned} (B_1/T^m)^d - (B_2/T^m)^d \in \pi_\infty^{m+1}\mathcal{O}_\infty &\Rightarrow \mathrm{v}_\infty(B_1^d - B_2^d) - \mathrm{v}_\infty(T^{md}) > m \\ &\Rightarrow \deg(B_1^d - B_2^d) < md - m = m(d-1)\,. \end{aligned}$$

On the other hand, if we let ζ run through the d-th roots of unity in some extension field of $\mathbb{F}_q$, we have (since $d < p$) that

$$\begin{aligned} \deg(B_1^d - B_2^d) &= \deg\left(\prod_\zeta B_1 - \zeta B_2\right) \\ &= \sum_\zeta \deg(B_1 - \zeta B_2) \\ &\geq \sum_{\zeta \neq 1} \deg(B_1 - \zeta B_2) = m(d-1)\,. \end{aligned}$$

Hence the cosets $(B/T^m)^d + \pi_\infty^{m+1}\mathcal{O}_\infty$ are disjoint. Since their total measure is $q^m \cdot q^{-m} = 1$, we see that their union must be $\mathrm{U}_\infty^{(1)} = 1 + \pi_\infty\mathcal{O}_\infty$. Therefore

$$\begin{aligned} q^{-m} \sideset{}{'}\sum_{\deg B = m} E_\infty(yB^d) &= \int_{1+\pi_\infty\mathcal{O}_\infty} E_\infty(yT^{md}u)\, du \\ &= E_\infty(yT^{md}) \int_{\pi_\infty\mathcal{O}_\infty} E_\infty(yT^{md}w)\, dw \\ &= \begin{cases} E_\infty(yT^{md}) & \text{if } \mathrm{v}_\infty(y) > md \\ 0 & \text{otherwise} \end{cases} \end{aligned}$$

where we have used the fact that $n(E_\infty) = 2$ and $\mathrm{v}_\infty(w) \geq 1$. This completes the proof. ∎

Let us apply Proposition 8.5 to the summation over B in Proposition 8.3. We have $m = n - j$; and because $t \in D_Q$, $y = t_\infty Q^d$ satisfies

$$\begin{aligned} v_\infty(y) &= v_\infty(t_\infty) + d v_\infty(Q) > n(d-1) + j - dj \\ &= (n-j)(d-1) = m(d-1)\,. \end{aligned}$$

Further if $v_\infty(y) > md$, then $v_\infty(t_\infty) > nd$ so that

$$\begin{aligned} v_\infty\Big(t_\infty T^{(n-j)d} Q^d - t_\infty T^{nd}\Big) &= v_\infty(t_\infty) + v_\infty(T^{nd}) + v_\infty\big((Q/T^j)^d - 1\big) \\ &> nd - nd + 1 = 1\,, \end{aligned}$$

which shows that $E_\infty\big(t_\infty T^{(n-j)d} Q^d\big) = E_\infty\big(t_\infty T^{nd}\big)$. The following evaluation is therefore a consequence of Proposition 8.5.

Corollary 8.6. *We have*

$$\sum_{\deg B = n-j}{}' \; E_\infty(t_\infty B^d Q^d) = \begin{cases} q^{n-j} E_\infty(t_\infty T^{nd}) & \textit{if } v_\infty(t_\infty) > nd \\ 0 & \textit{otherwise.} \end{cases}$$

Propositions 8.3 and 8.4 together with Corollary 8.6 yield

Proposition 8.7. *We have*

$$g(t) = \begin{cases} q^n E_\infty(t_\infty T^{nd}) \int_{\mathcal{O}} \tilde{E}(\tilde{t} z^d)\, dz & \textit{if } v_\infty(t_\infty) > nd \\ 0 & \textit{otherwise} \end{cases}$$

provided that $t \in D_Q$ for some Q with $\deg Q \le n$.

We may now evaluate the integral in (8.5) on the union of the major arcs in terms of the derivative $\theta(z; h, \mathcal{O}^s)$ of the map $h : \mathcal{O}^s \mapsto \mathcal{O}$ of (8.2). When $s > 2d$, this derivative exists as a continuous function on $\mathcal{O}$ by Theorem 4.38. Although $\theta(z; h, \mathcal{O}^s)$ is defined only for $z \in \mathcal{O}$, we can extend it to all of $\tilde{\mathbb{A}}_k$ by defining it to be zero on $\tilde{\mathbb{A}}_k - \mathcal{O}$. We have then

$$\hat{\theta}(\tilde{t}; h, \mathcal{O}^s) = \int_{\mathcal{O}} \theta(z; h, \mathcal{O}^s) E(-\tilde{t} z)\, dz$$

which equals the Fourier coefficient of $\theta(z; h, \mathcal{O}^s)$ as a function on the compact group $\mathcal{O}$ when $\tilde{t} = A/Q$. Let

$$D_{\text{MAJOR}} = \bigcup_{\deg Q \le n} D_Q \qquad \text{and} \qquad D_{\text{minor}} = \bigcup_{n < \deg Q \le n(d-1)} D_Q\,.$$

Theorem 8.8. *Assume $s > 2d$. Then*

$$\int_{D_{\text{MAJOR}}} g(\gamma_1\tilde{t})\, g(\gamma_2\tilde{t}) \cdots g(\gamma_s\tilde{t})\, E(-M\tilde{t})\; d\tilde{t} = q^{n(s-d)} \int_{\widetilde{D}_{\text{MAJOR}}} \hat{\theta}(\tilde{t}; h, \mathcal{O}^s)\, E(-M\tilde{t})\; d\tilde{t}. \tag{8.6}$$

Proof. For Q positive with $j = \deg Q \le n$, we conclude immediately from Proposition 8.7 that

$$\int_{D_Q} g(\gamma_1\tilde{t})\, g(\gamma_2\tilde{t}) \cdots g(\gamma_s\tilde{t})\, E(-M\tilde{t})\; d\tilde{t} = \int_{\widetilde{D}_Q} R(\tilde{t})\; d\tilde{t} \cdot \int_{\pi_\infty^{nd+1}\mathcal{O}_\infty} q^{ns} E_\infty\big((\gamma T^{nd} - M)t_\infty\big)\; dt_\infty,$$

where

$$R(\tilde{t}) = \int_{\mathcal{O}^s} \widetilde{E}\big((\gamma_1\tilde{t}_1^d + \gamma_2\tilde{t}_2^d + \cdots + \gamma_s\tilde{t}_s^d - M)\tilde{t}\big)\; d\tilde{t}_1 d\tilde{t}_2 \cdots d\tilde{t}_s$$

because the ∞-component of D_Q is $\pi_\infty^{n(d-1)+j+1}\mathcal{O}_\infty \supseteq \pi_\infty^{nd+1}\mathcal{O}_\infty$. Summing over all positive Q with $\deg Q \le n$, we get

$$q^{ns} \int_{\pi_\infty^{nd+1}\mathcal{O}_\infty} E_\infty\big((\gamma T^{nd} - M)t_\infty\big)\; dt_\infty \cdot \int_{\widetilde{D}_{\text{MAJOR}}} R(\tilde{t})\; d\tilde{t} \tag{8.7}$$

where γ is defined by (8.1). Since γ is the nd-th coefficient of M, we see that

$$\deg(\gamma T^{nd} - M) \le nd - 1 \Rightarrow \mathrm{v}_\infty\big((\gamma T^{nd} - M)t_\infty\big) \ge -(nd-1) + nd + 1 = 2$$

for all $t_\infty \in \pi_\infty^{nd+1}\mathcal{O}_\infty$. Therefore for $t_\infty \in \pi_\infty^{nd+1}\mathcal{O}_\infty$,

$$E_\infty\big((\gamma T^{nd} - M)t_\infty\big) = 1$$

and so the first integral in (8.7) evaluates to $meas(\pi_\infty^{nd+1}\mathcal{O}_\infty) = q^{-nd}$.

Now, by definition of the global derivative $\theta(z; h, \mathcal{O}^s)$,

$$R(\tilde{t}) = E(-M\tilde{t}) \int_{\mathcal{O}} \theta(z; h, \mathcal{O}^s) E(\tilde{t}z)\, dz = \hat{\theta}(\tilde{t}; h, \mathcal{O}^s) E(-M\tilde{t})$$

so that the second integral in (8.7) equals the integral on the right hand side of (8.6). ∎

When $d = 2$, $D = D_{\text{MAJOR}}$ because then D_{minor} is empty. Therefore Theorem 8.8 together with (8.5) provides an exact formula for $N_{2,s}(M)$. This exact formula was known to Carlitz in 1947 (cf. Exercise 8.1(2) below).

Exercises 8.1

1. In his analysis of the polynomial Waring problem, Webb (1973a) arrives at the following major term in the expression for $N_{d,s}(M)$:

$$q^{n(s-d)}\left[\sum_{\deg Q\le n}{}' q^{-js}\sum_{\substack{\deg A<j\\(A,Q)=1}} S(\gamma_1 A,Q)\ldots S(\gamma_s A,Q)\widetilde{E}(-MA/Q)\right]$$

   where $S(A,Q)=\sum_{\deg C<j}\widetilde{E}(C^dA/Q)$. This expression is valid for $s>d$. Obtain this result not by using Proposition 8.4, but rather by observing that the 'Q-toroid' $\mathcal{T}_Q$ (see (6.25)) is just

$$\bigcup_{\substack{\deg A<j\\(A,Q)=1}}(A/Q+\mathcal{O}_\infty)$$

   and the fact that for $z\in\mathcal{O}_\infty$, $\widetilde{E}(\tilde{t}z)=\widetilde{E}(Az/Q)$ if $\tilde{t}-A/Q\in\mathcal{O}$. Now evaluate

$$\int_{\mathcal{T}_Q} g(\gamma_1\tilde{t})\,g(\gamma_2\tilde{t})\cdots g(\gamma_s\tilde{t})\,E(-M\tilde{t})\,d\tilde{t},$$

   yielding Webb's expression. Since this expression is just the sum of the terms corresponding to the A/Q with $\deg Q\le n$ in the global singular series (4.24) for $\theta(z;h,\mathcal{O}^s)$, it can be analysed accordingly, giving a somewhat different route to a polynomial Waring theorem than the one we are following.

2. When $d=2$, the expression above for the major term of $N_{2,s}(M)$ is in fact equal to $N_{2,s}(M)$. Compare this expression for $N_{2,s}(M)$ with that given in Carlitz (1947a).

8.2 Minor arc analysis: Weyl's inequality for polynomials

Following the line of attack used by Hardy and Littlewood on the classical Waring problem, we now show that the magnitude of the integral in (8.5) on the union of the minor arcs grows at a rate which is strictly less than $q^{n(s-d)}$ (under suitable conditions on s). This will follow almost immediately from Theorem 8.11 below, which is a polynomial version of the famous Weyl inequality. In proving this result, we will need the following standard lemma.

Lemma 8.9. *For L a polynomial in $\mathbf{A}_q$, let $\tau(L)$ be the number of positive divisors of L. Then for every $\epsilon > 0$, we have*

$$\tau(L) \leq C_\epsilon |L|^\epsilon$$

where C_ϵ depends only on ϵ (and not on L or q).

This lemma is proved in exactly the same way as its classical analogue (see Theorem 315 of Hardy and Wright (1979) and Exercise 1).

Definition 8.10. *Let $f(z) = t_d z^d + t_{d-1} z^{d-1} + \ldots + t_0$ be a polynomial with coefficients in $\mathfrak{A}_k$. We put*

$$S_d(f, n) \stackrel{\text{def}}{=} \sum_{\deg A = n}{}' E(f(A)),$$

and we call $S_d(f, n)$ a Weyl sum.

Theorem 8.11. *Suppose that* $\text{char}(\mathbf{A}_q) = p > d$, *and suppose that the leading coefficient t_d of $f(z)$ satisfies $t_d \in D_Q$ where $n < \deg Q \leq n(d-1)$ (i.e. t_d lies in a minor arc). Then for every $\epsilon > 0$,*

$$|S_d(f, n)| = O\left(q^{n(1-1/2^{d-1}+\epsilon)}\right)$$

where the constant implied by the O-notation depends only on ϵ.

Proof. If $A \in \mathbf{A}_q$, we define the operator Δ_A on $f(z) \in \mathfrak{A}_k[z]$ by

$$\Delta_A f(z) \stackrel{\text{def}}{=} f(z + A) - f(z).$$

If $A_1, \ldots, A_\nu \in \mathbf{A}_q$, we write

$$\Delta_{A_1 \ldots A_\nu} \stackrel{\text{def}}{=} \Delta_{A_1} \circ \Delta_{A_2} \circ \cdots \circ \Delta_{A_\nu}$$

for the composite operator. Using this notation, we first show by induction that

$$|S_d(f, n)|^{2^\nu} \leq q^{n(2^\nu - \nu - 1)} \sum_{\substack{A_1, A_2, \ldots, A_\nu \\ \deg A_i < n}} S_{d-\nu}(\Delta_{A_\nu, A_{\nu-1}, \ldots, A_1} f, n) \qquad (8.8)$$

for $1 \leq \nu < d$. Suppose that $\nu = 1$. Then

$$\begin{aligned} |S_d(f, n)|^2 &= \sum_{\substack{A_1, A_2 \\ \deg A_i < n}} E(f(T^n + A_1) - f(T^n + A_2)) \\ &= \sum_{\deg A_2 < n} \sum_{\deg A_1 < n} E(f(T^n + A_1 + A_2) - f(T^n + A_2)) \\ &= \sum_{A_1} \sum_{A_2} E(\Delta_{A_1} f(T^n + A_2)) = \sum_{A_1} S_{d-1}(\Delta_{A_1} f, n) \end{aligned}$$

as desired. We now make the inductive step. Let us assume that (8.8) holds for some $1 \leq \nu < d - 1$. We square both sides and apply Cauchy's inequality on the right, using the fact that there are q^n polynomials of degree less than n and also using the result just proved for $\nu = 1$. We obtain

$$\begin{aligned} |S_d(f,n)|^{2^{\nu+1}} &\le q^{n(2^{\nu+1}-2\nu-2)} \left(\sum_{\substack{A_1,\dots,A_\nu \\ \deg A_i<n}} S_{d-\nu}(\Delta_{A_\nu \dots A_1} f, n) \right)^2 \\ &\le q^{n(2^{\nu+1}-2\nu-2)} q^{n\nu} \sum_{A_1,\dots,A_\nu} |S_{d-\nu}(\Delta_{A_\nu \dots A_1} f, n)|^2 \\ &= q^{n(2^{\nu+1}-(\nu+1)-1)} \sum_{A_1,\dots,A_{\nu+1}} S_{d-\nu-1}(\Delta_{A_{\nu+1}\dots A_1} f, n) \end{aligned}$$

completing the induction and establishing (8.8).

Now put $\nu = d-1$ in (8.8). Since f is a polynomial of degree d, we see easily by induction on d that

$$\Delta_{A_{d-1}\dots A_1} f(z) = d! A_1 A_2 \dots A_{d-1} t_d z + \phi(A_1, \dots, A_{d-1})$$

where ϕ is some polynomial in A_1 through A_{d-1}. Therefore

$$\left|S_1(\Delta_{A_{d-1}\dots A_1} f, n)\right| = \left| \sum_{\deg A=n}{}' E(d!A_1, \dots, A_{d-1} t_d A) \right| . \tag{8.9}$$

There are clearly less than $dq^{n(d-2)}$ $(d-1)$-tuples $(A_1, A_2, \dots, A_{d-1})$ with at least one zero entry. For each of these, the sum on the right above is q^n. Therefore (8.8) with $\nu = d-1$ implies

$$|S_d(f,n)|^{2^{d-1}} \le dq^{n(2^{d-1}-1)} + q^{n(2^{d-1}-d)} T_1 \tag{8.10}$$

where

$$T_1 \overset{\text{def}}{=} \sum_{\substack{A_1,\dots,A_{d-1} \\ A_i \ne 0}} \left| \sum_{\deg A=n}{}' E(A_1, \dots, A_{d-1} t_d A) \right| \tag{8.11}$$

because, since $p > d$, $d!A_1$ runs with A_1 through all polynomials of degree strictly less than n. We must now estimate T_1.

First, we need to count the number $\eta(L)$ of non-zero solutions of the equation $L = A_1 A_2 \dots A_{d-1}$, $L \ne 0$. Let $\tau(L)$ be the number of positive divisors of L. By picking arbitrary positive divisors and coefficients in the first $d-2$ slots, we see that

$$\eta(L) \le (q \cdot \tau(L))^{d-2}.$$

Combining this inequality with Lemma 8.9, we obtain

$$\begin{aligned} T_1 &= \sum_{\substack{L \neq 0 \\ \deg L \le (n-1)(d-1)}} \eta(L) \left| \sum_{\deg A = n}{}' E(Lt_d A) \right| \\ &\le \sum_{\substack{L \neq 0 \\ \deg L \le (n-1)(d-1)}} q^{d-2} C_\epsilon^{d-2} q^{\deg L \cdot \epsilon(d-2)} \left| \sum_{\deg A = n}{}' E(Lt_d A) \right|, \end{aligned}$$

so that

$$T_1 \le q^{d-2} C_\epsilon^{d-2} q^{\epsilon n(d-1)(d-2)} \sum_{\substack{L \\ \deg L \le (n-1)(d-1)}} \left| \sum_{\deg A = n}{}' E(Lt_d A) \right|.$$

We must now estimate the double summation on the right above. First we observe that since $t_d \in D_Q$ with $n < \deg Q \le n(d-1)$, we have

$$v_\infty(Lt_d A) \ge n(d-1) + (n+1) + 1 - (n-1)(d-1) - n = d+1 \ge 2$$

so that E_∞ is trivial on $Lt_d A$. Let $j = \deg Q$. Since $Q = \mathfrak{n}(t_d)$, the finite part of the conductor of t_d, there exists a polynomial B prime to Q such that $\tilde{t}_d - B/Q \in \mathcal{O}$. We have then $\widetilde{E}(S\tilde{t}_d) = \widetilde{E}(SB/Q)$ for every $S \in \mathbf{A}_q$. Therefore

$$LB \equiv C \pmod{Q} \Rightarrow E(Lt_d A) = \widetilde{E}(L\tilde{t}_d A) = \widetilde{E}(CA/Q).$$

Now, for fixed C with $\deg C < j$, there are at most $(q^{(n-1)(d-1)-j} + 1)$ polynomials L of degree $(n-1)(d-1)$ or less such that LB is congruent to C modulo Q. Thus

$$T_2 \stackrel{\text{def}}{=} \sum_{\deg L \le (n-1)(d-1)} \left| \sum_{\deg A = n}{}' \widetilde{E}(L\tilde{t}_d A) \right|$$

$$\leq \left(q^{(n-1)(d-1)-j}+1\right)\sum_{\deg C<j}\left|\sum_{\deg A=n}{}' \widetilde{E}(CA/Q)\right| .$$

Now since E is trivial on k (see Proposition 4.8), we have $\widetilde{E}(CA/Q) = E_\infty(-CA/Q)$. Because $v_\infty(C/A) > 0$, we may apply Proposition 8.5 with $d = 1$ to conclude that

$$\left|\sum_{\deg A=n}{}' E_\infty(-CA/Q)\right| = \begin{cases} q^n & \text{if } v_\infty(C/Q) > n \\ 0 & \text{otherwise.}\end{cases}$$

But $v_\infty(C/Q) > n \iff \deg C < j - n$, and there are q^{j-n} polynomials C which satisfy this inequality. Hence we arrive at

$$\sum_{\deg C<j}\left|\sum_{\deg A=n}{}' E_\infty(-CA/Q)\right| = q^n(q^{j-n}) = q^j$$

which implies $T_2 \leq q^{(n-1)(d-1)} + q^j \leq 2q^{n(d-1)}$ so that

$$T_1 \leq q^{d-2}C_\epsilon^{d-2}q^{\epsilon n(d-1)(d-2)}T_2 \leq 2q^{n(d-1)+d-2}C_\epsilon^{d-2}q^{\epsilon n(d-1)(d-2)} .$$

Substituting from this last inequality into (8.10), we get

$$\begin{aligned} |S_d(f,n)|^{2^{d-1}} &\leq q^{n(2^{d-1}-1)}\left[d + q^{d-2}C_\epsilon^{d-2}q^{\epsilon n(d-1)(d-2)}\right] \\ &\leq q^{n(2^{d-1}-1+\epsilon 2^{d-1})}\left[d + q^{d-2}C_\epsilon^{d-2}\right] \end{aligned}$$

because $2^{d-1} \geq (d-1)(d-2)$ for all $d \geq 1$. Taking the 2^{d-1}-st root of both sides, we finally obtain

$$|S_d(f,n)| \leq \left[d + q^{d-2}C_\epsilon^{d-2}\right]^{2^{1-d}} q^{n(1-1/2^{d-1}+\epsilon)} .$$

The reader can easily check that the first factor on the right is a bounded function of d. This completes the proof of Weyl's inequality. ■

We may now prove our minor arc bound.

Theorem 8.12. *If $s \geq d2^d$ and $0 < \delta < d$, then*

$$\left|\int_{D_{\text{minor}}} g(\gamma_1 t)\cdots g(\gamma_s t)E(-Mt)\,dt\right| = O\left(q^{n(s-d-\delta)}\right) , \tag{8.12}$$

where the implied constant depends on d, s, and δ but not on q or n.

Proof. For $1 \le i \le s$, each function $g(\gamma_i t)$ is a Weyl sum with $f_i(u) = \gamma_i t u^d$ (see (8.4) and Definition 8.10). Hence by Theorem 8.11, for any $\epsilon > 0$,

$$\begin{aligned}\left|\int_{D_{\text{minor}}} g(\gamma_1 t)\cdots g(\gamma_s t)E(-Mt)\,dt\right| &\le \int_{D_{\text{minor}}} |S_d(f_1,n)|\cdots|S_d(f_s,n)|\,dt \\ &= O\left(q^{ns(1-1/2^{d-1}+\epsilon)}\right)\end{aligned}$$

since $meas(D_{\text{minor}}) \le 1$. Now let $s \ge d2^d$ and $0 < \delta < d$. Setting $\epsilon = 1/2^d - \delta/d2^d > 0$, we have

$$1 - 1/2^{d-1} + \epsilon = 1 - 1/2^d - \delta/d2^d$$

so that

$$s(1 - 1/2^{d-1} + \epsilon) \le s - d - \delta$$

as required. ∎

We have proved that the integral in (8.5) grows more slowly on the minor arcs than on the major arcs.

Exercises 8.2

1. We outline below a proof of Lemma 8.9:

   (a) Since both $\tau(L)$ and $|L|$ are multiplicative functions of L, write $\tau(L)/|L|^\epsilon$ as a product over its prime divisors, and separate the product into two parts: those with $|P| \ge 2^{1/\epsilon}$ and those with $|P| < 2^{1/\epsilon}$.

   (b) Show that if $|P| \ge 2^{1/\epsilon}$, then $\tau(P^e)/|P^e|^\epsilon \le 1$ for all e, so that that part of the product can be omitted.

   (c) Show that for any fixed P and ϵ, $\tau(P^e)/|P^e|^\epsilon$ possesses a maximum value over all $e \ge 1$, so set $M_\epsilon = \max_{e\ge1}\{(e+1)/2^{e\epsilon}\}$.

   (d) Show finally that

$$\begin{aligned}\tau(L)/|L|^\epsilon &\le \prod_{|P|<2^{1/\epsilon}} M_\epsilon \\ &\le \prod_{1\le d\le \log 2/\epsilon \log q} M_\epsilon^{q^d} \\ &\le M_\epsilon^{(2^{1/\epsilon}/\epsilon)} = C_\epsilon,\end{aligned}$$

   as required.

2. Prove the following analogue of Hua's inequality for polynomials:

Theorem 8.13. *If $g(t)$ is as in (8.4), then for every $\epsilon > 0$,*

$$\int_D |g(t)|^{2^d} \, dt = O\left(q^{n(2^d - d + \epsilon)}\right),$$

where the implied constant depends on d, ϵ, and *q, but not on n.*

The proof of this theorem is an essentially direct translation of the classical result. Using for example Ayoub (1963, page 242), work through the details of this translation. You will need to use our inequality (8.8) with $\nu = m - 1$. The dependence on q arises from the fact that the polynomials $A_1, \ldots, A_{m-1}$ are not positive, and so in applying Lemma 8.9 to the equation

$$A_1 A_2 \ldots A_{m-1} \phi(A) = W,$$

there are $(q-1)\tau(W)$ choices for each A_i. One obtains finally that for $1 \leq m \leq d$,

$$|g(t)|^{2^m} = O\left((q-1)^{m-2} q^{n(2^m - m + \epsilon)}\right)$$

(where the implied constant depends only on d and ϵ). Our stated result follows.

3. Combine Weyl's and Hua's inequalities to obtain a result similar to Theorem 8.12:

Theorem 8.14. *If $s \geq 2^d + 1$, then there exists a $\delta > 0$ such that*

$$\left| \int_{D_{\text{minor}}} g(\gamma_1 t) g(\gamma_2 t) \cdot g(\gamma_s t) E(-Mt) \, dt \right| = O\left(q^{n(s - d - \delta)}\right),$$

where the implied constant depends on d, s, δ, and q, but not on n.

8.3 The Waring singular series and the polynomial Waring theorem

Throughout this section, we assume that $s > 2d$. In our attempt to establish that $N_{d,s}(M) > 0$ for sufficiently large s, it remains only to approximate the integral

$$I(M) \stackrel{\text{def}}{=} \int_{\widetilde{D}_{\text{MAJOR}}} \hat{\theta}(\tilde{t}; h, \mathcal{O}^s) E(-M\tilde{t})\, d\tilde{t}$$

of Theorem 8.8. Applying formal Fourier inversion (cf. Proposition 4.10), we obtain

$$\theta(M; h, \mathcal{O}^s) = \int_{\tilde{\mathbb{A}}_k} \hat{\theta}(\tilde{t}; h, \mathcal{O}^s) E(-M\tilde{t})\, d\tilde{t}$$

so that at least formally

$$I(M) = \theta(M; h, \mathcal{O}^s) - I_n(M) \tag{8.13}$$

where

$$I_n(M) \stackrel{\text{def}}{=} \int_{\tilde{\mathbb{A}}_k - \widetilde{D}_{\text{MAJOR}}} \hat{\theta}(\tilde{t}; h, \mathcal{O}^s) E(-M\tilde{t})\, d\tilde{t}.$$

We will estimate the magnitude of $I_n(M)$, proving at the same time that this integral is convergent. It will then follow that $\hat{\theta}(\tilde{t}; h, \mathcal{O}^s) \in L^1(\tilde{\mathbb{A}}_k)$, justifying the Fourier inversion and validating eqn (8.13). Since $\tilde{\mathbb{A}}_k$ is the disjoint union of its subsets $\widetilde{D}_Q$, Q running over all positive polynomials in $\mathbf{A}_q$, we have

$$\begin{aligned} |I_n(M)| &\le \sum_{\deg Q > n} \int_{\widetilde{D}_Q} |\hat{\theta}(\tilde{t}; h, \mathcal{O}^s)|\, d\tilde{t} \\ &\le \sum_{j=n+1}^{\infty} \sum_{\deg Q = j} meas(\widetilde{D}_Q) \cdot \sup_{\tilde{t} \in \widetilde{D}_Q} \left|\hat{\theta}(\tilde{t}; h, \mathcal{O}^s)\right| . \end{aligned} \tag{8.14}$$

By (3.23) and the remarks preceding Theorem 3.13, the local Fourier transform $\hat{\theta}_P(\tilde{t}_P; h_P, \mathcal{O}_P^s)$ satisfies the inequality

$$|\hat{\theta}_P(\tilde{t}_P; h_P, \mathcal{O}_P^s)| \le d^s \cdot |P|^{-v_P(Q)s/d}$$

for every P dividing $Q = \tilde{n}(\tilde{t})$. Therefore, the global Fourier transform satisfies

$$|\hat{\theta}(\tilde{t}; h, \mathcal{O}^s)| \le d^{s\omega(Q)} |Q|^{-s/d} \tag{8.15}$$

where $\omega(Q)$ is the number of distinct positive irreducibles which divide Q. Choose ϵ so that

$$0 < \epsilon < \frac{(s/d) - 2}{s \log_2 d} \tag{8.16}$$

and put

$$\mu \stackrel{\text{def}}{=} s/d - 2 - \epsilon s \log_q d \geq \mu_2 \stackrel{\text{def}}{=} s/d - 2 - \epsilon s \log_2 d\,.$$

Let $N(\epsilon)$ be the quantity associated to ϵ by Lemma 4.19. If $j = \deg Q$, then (8.15) yields

$$\sup_{\tilde{t} \in \widetilde{D}_Q} |\hat{\theta}(\tilde{t}; h, \mathcal{O}^s)| \leq d^{sN(\epsilon)} d^{s\epsilon j} q^{-js/d} = d^{sN(\epsilon)} q^{-j(s/d - s\epsilon \log_q d)} \tag{8.17}$$

because if $j < N(\epsilon)$, then $d^{s\omega(Q)} \leq d^{sN(\epsilon)}$ whereas if $j \geq N(\epsilon)$, then $d^{s\omega(Q)} \leq d^{s\epsilon j}$.

Combining (8.14) and (8.17) with the fact that $meas(\widetilde{D}_Q) = \Phi(Q) \leq q^j$ and the fact that there are q^j positive polynomials Q of degree j, we obtain

$$\begin{aligned} I_n(M) &\leq \sum_{j=n+1}^{\infty} \sum_{\deg Q = j} q^j d^{sN(\epsilon)} q^{-j(s/d - s\epsilon \log_q d)} \\ &= d^{sN(\epsilon)} \sum_{j=n+1}^{\infty} q^{-j\mu} \leq d^{sN(\epsilon)} \sum_{j=n+1}^{\infty} q^{-j\mu_2} \\ &= \left(\frac{d^{sN(\epsilon)}}{q^{\mu_2} - 1} \right) q^{-n\mu_2}\,. \end{aligned} \tag{8.18}$$

This inequality together with (8.13) shows that

$$I(M) = \theta(M; h, \mathcal{O}^s) + O\left(q^{-n\mu_2}\right)\,,$$

the implied constant depending upon s and d but not upon q or n. Combining this asymptotic formula with Theorem 8.8, we get

$$\int_{D_{\text{MAJOR}}} g(\gamma_1 t) \cdots g(\gamma_s t) E(-Mt)\, dt$$
$$= q^{n(s-d)} \theta(M; h, \mathcal{O}^s) + O\left(q^{n(s-d-\mu_2)}\right)\,. \tag{8.19}$$

We may now complete the proof of the asymptotic Waring theorem (Theorem 8.1). If we choose the parameter δ of Theorem 8.12 to equal $\min(\mu_2, d-1)$, then the asymptotic formula (8.3) follows immediately from (8.5), (8.12), and (8.19).

The asymptotic formula (8.3) is of little use unless we can demonstrate that the global Radon–Nikodym derivative $\theta(z; h, \mathcal{O}^s)$ on $\mathcal{O}$ is bounded below by a positive constant which is independent of q. For $s \geq d2^d > d^2$, Theorem 4.18 and Theorem 3.14 show that $\theta(z; h, \mathcal{O}^s)$ does not vanish on $\mathcal{O}$. Since $\mathcal{O}$ is compact, there is a constant $\kappa_q > 0$ such that $\theta(z; h, \mathcal{O}^s) \geq \kappa_q$. It suffices, therefore, to find a constant $\kappa > 0$ such that $\kappa_q \geq \kappa$ for all sufficiently large q. Now by Theorem 4.20, $\theta(z; h, \mathcal{O}^s)$ is represented by its Fourier series. By the technique of (8.18), we have then

$$|\theta(z; h, \mathcal{O}^s) - 1| \leq \sum_{j=1}^{\infty} d^{sN(\epsilon)} q^{-j\mu_2} \leq \frac{d^{sN(\epsilon)}}{q^{\mu_2} - 1}.$$

Thus, $\theta(z; h, \mathcal{O}^s)$ is close to one for all sufficiently large q. This together with Theorem 8.1 gives us our main result.

Theorem 8.15. *Let $d \geq 2$ be any integral exponent and assume that $p = \operatorname{char}(\mathbf{A}_q) > d$. If $s \geq d2^d$, then for every $M \in \mathbf{A}_q$ with $n = \{\deg M/d\}$ sufficiently large and every s-tuple $(\gamma_1, \ldots, \gamma_s)$, $\gamma_i \in \mathbf{F}_q^{\times}$, with $\sum_{i=1}^{s} \gamma_i =$ coefficient of T^{nd} in M, there exist positive polynomials $X_1, X_2, \ldots, X_s$ in $\mathbf{A}_q$, each of degree n, such that*

$$M = \gamma_1 X_1^d + \gamma_2 X_2^d + \cdots + \gamma_s X_s^d.$$

Further if q is sufficiently large, then this same conclusion holds for all $M \in \mathbf{A}_q$.

The polynomial Waring theorem (Theorem 1.9) is an easy corollary of Theorem 8.15 together with Lemma 1.12. We have only to show that for $s = d2^d$ and $p = \operatorname{char}(\mathbf{F}_q) > d \geq 2$, we can write any $\gamma \in \mathbf{F}_q$ as the sum of s *non-zero* elements of $\mathbf{F}_q$. This is certainly the case if $\gamma = 0$ because $0 = m \cdot 1 + m \cdot (-1)$ where $m = d2^{d-1}$. If $\gamma \neq 0$, then $\gamma = s \cdot (s^{-1}\gamma)$. Since every element of $\mathbf{F}_q$ is a sum of not more than d d-th powers by Lemma 1.5, the polynomial Waring theorem follows.

Exercises 8.3

1. Use Theorem 8.14 of Exercise 8.2(3) to show that the conclusion of Theorem 8.15 holds for $s = 2^d + 1$ and all $M \in \mathbf{A}_q$ with $\deg M \geq C_q$, where $C_q > 0$ is a constant which may depend upon q.

2. Suppose $\gamma \neq 0$ in (8.1). Using the technique of §3.3, show that the Radon–Nikodym derivative of the map

$$h : \left(\mathbf{U}_\infty^{(1)}\right)^s \longrightarrow \gamma \mathbf{U}_\infty^{(1)}$$

defined by the polynomial

$$h(x_1, x_2, \ldots, x_s) = \gamma_1 x_1^d + \gamma_2 x_2^d + \cdots + \gamma_s x_s^d$$

is the constant function $\theta(z) = 1$. If $\gamma = 0$, show that the Radon–Nikodym derivative of the map

$$h : \left(\mathbf{U}_\infty^{(1)}\right)^s \longrightarrow T^{-1}\mathcal{O}_\infty$$

is also unity.

3. Put

$$V = \left(T^n \mathbf{U}_\infty^{(1)}\right)^s,$$

a space of measure q^{sd}. Show that the factor $q^{n(s-d)}$ in the asymptotic formula (8.3) is the (constant) Radon–Nikodym derivative of the map $h : V \to W_\gamma$, where

$$W_\gamma = \begin{cases} T^{nd}\gamma \mathbf{U}_\infty^{(1)} & \text{if } \gamma \neq 0 \\ T^{nd-1}\mathcal{O}_\infty & \text{if } \gamma = 0. \end{cases}$$

Appendix A
A complete solution to the 3-primes problem

Theorem 7.11, the asymptotic theorem, is an existence theorem which guarantees, for any fixed degree $r \geq 5$, some field size q_r such that if $q \geq q_r$, then every (odd) positive polynomial M of degree r over $\mathbf{F}_q$ possesses representations as a sum of three positive irreducibles. An obvious numerical question is: what can we say about the numbers q_r? We can easily use the computer to obtain upper bounds for these numbers by starting with the inequality (7.32). However, in order to obtain bounds as sharp as possible, we shall develop a lower bound $S(q,r)$ for the singular series $\theta(M;h,V)$ which is an increasing function of both q and r and which is slightly sharper than the one obtained in §5.4. Then, for any fixed $r \geq 5$, we find the smallest q such that

$$\frac{q^{r/4}}{r(q-1)(r-1)} > \frac{C(q,r)}{S(q,r)} \tag{A.1}$$

where $C(q,r) = C_1 + C_2 + C_3 + C_4$ is exactly the decreasing function of q and r in Theorem 7.10.

To get our lower bound on $\theta(M;h,V)$, we start with the representation of $\theta(M;h,V)$ given in Theorem 4.15 for the case $\tilde{t} = M$. Multiplying and dividing (4.27) by $\prod_{P|M}\left(1+1/(|P|-1)^3\right)$ and simplifying, we obtain

$$\theta(M;h,V) = \prod_{\text{all } P}\left(1+\frac{1}{(|P|-1)^3}\right)\prod_{P|M}\left(1-\frac{1}{|P|^2-3|P|+3}\right). \tag{A.2}$$

From the first (i.e. infinite) product in (A.2), we make use of the factors involving linear, quadratic, and cubic irreducibles, since for quartic and higher degree polynomials the factors are very close to unity even for small q. Since there are q, $(q^2-q)/2$, and $(q^3-q)/3$ irreducible polynomials of degree 1, 2, and 3 respectively, we define

$$S_1(q) = \left(1+\frac{1}{(q-1)^3}\right)^q\left(1+\frac{1}{(q^2-1)^3}\right)^{\frac{q^2-q}{2}}\left(1+\frac{1}{(q^3-1)^3}\right)^{\frac{q^3-q}{3}}.$$

For the second (i.e. finite) product in (A.2), we must take every term to obtain a lower bound. Since $q^{2x} - 3q^x + 3$ is an increasing function of x for all $x \geq 1$, we see that among polynomials M of degree r, the minimum $\theta(M; h, V)$ will occur for polynomials that have the maximum number of distinct irreducible factors of as low degree as possible. For example, if $r \leq q$, then the minimum $\theta(M; h, V)$ will result from an M which has r distinct linear irreducible factors. Using the same principle as above, and letting

$$\alpha_q = \left(1 - \frac{1}{q^2 - 3q + 3}\right)^q$$

we define, for $q \neq 2$:

$$S_2(q,r) = \begin{cases} (1 - \frac{1}{q^2-3q+3})^r & \text{if } r \leq q \\ \alpha_q \cdot (1 - \frac{1}{q^4-3q^2+3})^{\frac{r-q}{2}} & \text{if } q < r \leq q^2 \\ \alpha_q \cdot (1 - \frac{1}{q^4-3q^2+3})^{\frac{q^2-q}{2}} (1 - \frac{1}{q^6-3q^3+3})^{\frac{r-q^2}{3}} & \text{if } r > q^2 . \end{cases}$$

It is easy to check that S_2 gives *the* minimum value for the finite product in (A.2) for any given r and q provided that $r \leq q^2$, and otherwise gives a very good approximation since higher order factors do not significantly alter the product.

We note that for $q = 2$, we must omit all even polynomials, and hence the linear terms are missing. In this case, making use of quartics and quintics, we let

$$\begin{aligned} S_2(2,r) &= \left(1 - \frac{1}{2^4 - 3 \cdot 2^2 + 3}\right)^{\frac{2^2-2}{2}} \cdots \left(1 - \frac{1}{2^{10} - 3 \cdot 2^5 + 3}\right)^{\frac{r-20}{5}} \\ &= (6/7)(42/43)^2(210/211)^3(930/931)^{\frac{r-20}{5}} \end{aligned}$$

where we have assumed that $r \geq 20$. The reader may check the details of this definition.

We have then that $S_1(q) \cdot S_2(q) \leq \theta(M; h, V)$ for all odd M of degree r over $\mathbf{F}_q$, and we define

$$S(q,r) = S_1(q) \cdot S_2(q,r).$$

We now turn to the computer, key in various values of q and r, and see for what pairs of values the inequality (A.1) is true. The results of these computations are summarized in Table A.1.

Table A.1 is the first step in the proof of the polynomial 3-primes theorem (Theorem 1.23), which we restate here:

Table A.1.

For odd polynomials of degree $r =$	the asymptotic theorem applies provided $q \geq$
5	2 352 841
6	2999
7	311
8	97
9	47
10	29
11	23
12	16
13	13
14	11
15	9
16	8
17–20	7
21–24	5
25–33	4
34–41	3
42 and up	2

Theorem A.1. *Every odd positive polynomial M of degree $r \geq 2$ over every finite field can be written as a sum of three positive irreducible polynomials (one of degree r and the others of degree less than r) except for the case $M = T^2 + \alpha$ when q is even.*

This result is the 'best possible', for it is easy to show that there exist even polynomials of arbitrarily high degree which cannot be written as a sum of three irreducibles (Exercise 7.3(1)), and likewise that when q is a power of 2, $M = T^2 + \alpha$ cannot be so written (Effinger 1988).

A proof of the polynomial 3-primes theorem depends then on covering the cases not included in Table A.1. The following result takes care of the 'low degree cases' (Effinger 1983, 1988):

Theorem A.2. *Every odd positive polynomial of degree $r = 2, 3, 4,$ or 5 over every finite field $\mathbf{F}_q$ is a sum of three positive irreducibles except for the case $M = T^2 + \alpha$, q even. Every positive polynomial of degree $r = 6$ can be so written provided $q \geq 19$ and every positive polynomial of degree $r = 7$ can be so written provided $q \geq 211$ but $q \neq 256$.*

Theorem A.2 reduces our conjecture to a computation which is finite but, it turns out, not tractable. For this the following two lemmas, especially the latter, are essential (see Effinger (1983, 1988) for the first and Theorem 9.3 of Hayes (1965) for the second).

Lemma A.3. *If q and r are relatively prime, then it suffices to check for* '3-primes representations' *only of polynomials with first coefficient* 0 *and second coefficient* 0, 1, *or, for q odd, some fixed quadratic non-residue.*

Lemma A.4. *Among positive polynomials of degree r over $\mathbb{F}_q$, there exist irreducible polynomials with every possible choice of first s coefficients provided that*

$$r/2 > s + \log_q(s+1).$$

If M is a monic polynomial of degree r for which we seek a 3-primes representation, Lemma A.4 guarantees us a monic irreducible polynomial P_1 of degree r such that $M - P_1$ is monic of degree a little more than half of r. Therefore, we need a 2-primes representation of $M - P_1$. Now, employing Lemma A.3, we find exactly 85 combinations of q and r which require checking (e.g. $q = 256$ $r = 5$, $q = 199$ $r = 4, \ldots, q = 2$ $r = 25$). Since the computation is still quite large, a powerful machine is needed. In 1989, the IBM 3090 Supercomputer at the Cornell National Supercomputing Facility in Ithaca, New York, was programmed to check these cases, and the results were all affirmative. Thus, the polynomial 3-primes theorem holds without exception. See Effinger (1991) for the details of the argument which has been outlined here.

For an overview of the complete theorem and its proof, the reader may consult Effinger and Hayes (1991). That paper contains a table which differs from Table A.1 above at the $r = 5$ and $r = 12$ entries. These two errors resulted from a slight miscalculation. The values in Table A.1 are correct.

Bibliography

1. Artin, E. (1924). Quadratische Körper im gebiet der höhern Kongruenzen I, II. *Mathematische Zeitschrift*, **19**, 153–246 = (1965). *Complete papers of Emil Artin*, pp.1–156. Addison-Wesley, Reading, MA.

2. Artin, E. (1957). *Geometric algebra.* Interscience Publishers, New York.

3. Ayoub, R. (1963). *An introduction to the analytic theory of numbers.* American Mathematical Society, Providence, RI.

4. Birch, B.J. (1961). Waring's problem for algebraic number fields. *Proceedings of the Cambridge Philosophical Society*, **57**, 449–59.

5. Birch, B.J. (1964). Waring's problem for p-adic number fields. *Acta Arithmetica,* **9**, 169–76.

6. Bombieri, E. (1974) Counting points on curves over finite fields (d'après S.A. Stepanov). In *Séminaire Bourbaki 1972–73*, Lecture Notes in Mathematics 383, pp.234–41. Springer-Verlag, New York.

7. Borevich, Z.I. and Shafarevich, I.R. (1966). *Number theory.* Academic Press, New York.

8. Car, M. (1971a). Le problème de Waring pour l'anneau des polynômes sur un corps fini. *Comptes Rendus de l'Académie des Sciences Paris, Series A–B*, **273**, A141–4.

9. Car, M. (1971b). Le problème de Goldbach pour l'anneau des polynômes sur un corps fini. *Comptes Rendus de l'Académie des Sciences Paris, Series A–B*, **273**, A201–4.

10. Car, M. (1972). *Arithmétique additive dans l'anneau des polynômes à une indéterminée sur un corps fini.* Thèses. Université de Provence.

11. Car, M. (1974). *Un problème d'arithmétique additive dans l'anneau des polynômes sur un corps fini.* Marseille-Luminy.

12. Car, M. (1981). Sommes de carrés et d'irréductibles dans $\mathbf{F}_q[x]$. *Annales Faculté des Sciences Toulouse*, **3**, 129–66.

13. Car, M. (1983). *Sommes de carrés dans* $\mathbf{F}_q[x]$. Dissertationes Mathematicae (Rozprawy Matematyczne), No. 225. Mathematical Institute of the Polish Academy of Sciences, Warsaw.

14. Car, M. (1984). Sommes de puissances et d'irréductibles dans $\mathbf{F}_q[X]$. *Acta Arithmetica*, **44**, 7–34.

15. Carlitz, L. (1932). The arithmetic of polynomials in a Galois field. *American Journal of Mathematics*, **54**, 39–50.

16. Carlitz, L. (1933). On representations of polynomials in a Galois field as a sum of an even number of squares. *Transactions of the American Mathematical Society*, **35**, 397–410.

17. Carlitz, L. (1935). On representations of polynomials in a Galois field as a sum of an odd number of squares. *Duke Mathematical Journal*, **1**, 298–315.

18. Carlitz, L. (1937). Sums of squares of polynomials. *Duke Mathematical Journal*, **3**, 1–7.

19. Carlitz, L. (1942). Some topics in the arithmetic of polynomials. *Bulletin of the American Mathematical Society*, **48**, 679–91.

20. Carlitz, L. (1947a). The singular series for sums of squares of polynomials. *Duke Mathematical Journal*, **14**, 1105–20.

21. Carlitz, L. (1947b). Representations of arithmetic functions in $GF[p^n, x]$ I. *Duke Mathematical Journal*, **14**, 1121–37.

22. Carlitz, L. (1948). Representations of arithmetic functions in $GF[p^n, x]$ II. *Duke Mathematical Journal*, **15**, 795–801.

23. Carlitz, L. (1952). Theorem of Dickson on irreducible polynomials. *Proceedings of the American Mathematical Society*, **3**, 693–700.

24. Carlitz, L. (1975). A note on sums of three squares in $GF[q, x]$. *Mathematics Magazine*, **48**, 109–10.

25. Cassels, J.W.S. (1967). Global fields. In *Algebraic number theory* (ed. J.W.S. Cassels and A. Frölich), pp.42–84. Thompson Book Company, Washington, DC.

26. Chen, J. (1973). On the representations of a large even integer as a sum of a prime and the product of at most two primes. *Scientia Sinica*, **16**, 157–76.

27. Cherly, J. (1977). Addition theorems in $\mathbb{F}_q[x]$. *Journal für die reine und angewandte Mathematik*, **293/294**, 223–7.

28. Cherly, J. (1978). A lower bound theorem in $\mathbb{F}_q[x]$. *Journal für die reine und angewandte Mathematik*, **303/304**, 253–64.

29. Cherly, J. (1989). *Sommes d'exponentielles cubiques dans l'anneau des polynômes en une variable sur le corps à deux éléments, et application au problème de Waring.* Thèses. Université de Bordeaux I.

30. Chevalley, C. (1951). *Introduction to the theory of algebraic functions of one variable.* American Mathematical Society, Providence, RI.

31. Cohen, E. (1947a). Sums of an even number of squares on $GF[p^n, x]$. *Duke Mathematical Journal*, **14**, 251–67.

32. Cohen, E. (1947b). Sums of an even number of squares on $GF[p^n, x]$, II. *Duke Mathematical Journal*, **14**, 543–57.

33. Cohen, E. (1948). Sums of an odd number of squares on $GF[p^n, x]$. *Duke Mathematical Journal*, **15**, 501–11.

34. Cohen, E. (1951). Sums of products of polynomials in a Galois field. *Duke Mathematical Journal*, **18**, 425–30.

35. Cohen, E. (1952). Arithmetic functions of polynomials. *Proceedings of the American Mathematical Society*, **3**, 352–8.

36. Davenport, H. (1962). *Analytic methods for Diophantine equations and approximations.* Ann Arbor Publishers, Ann Arbor, MI.

37. Dedekind, R. (1857). Abriss einer Theorie der höheren Congruenzen in Bezug auf einer reellen Primzahl–Modulus. *Journal für die reine und angewandte Mathematik*, **54**, 1–26.

38. Descombes, R. (1966–7). Partitions en trois polynômes irréducibles sur un corps fini. In *Séminaire Delange–Pisot–Poitou (Théorie des nombres)*, n° 12.

39. Dress, F. (1962–3). Fonctions arithmétiques sur l'anneau des polynômes à coefficients dans un corps fini. In *Séminaire Delange–Pisot–Poitou (Théorie des nombres)*, n° 13.

40. Effinger, G.W. (1983). A Goldbach theorem for polynomials of low degree over odd finite fields. *Acta Arithmetica*, **42**, 329–65.

41. Effinger, G.W. (1988). A Goldbach 3-primes theorem for polynomials of low degree over finite fields of characteristic 2. *Journal of Number Theory*, **29**, 345–63.

42. Effinger, G.W. (1991). The polynomial 3-primes conjecture. In *Computer assisted analysis and modeling on the IBM 3090.* MIT Press, Cambridge, MA.

43. Effinger, G.W. and Hayes, D.R. (1991). A complete solution to the polynomial 3-primes problem. *Bulletin of the American Mathematical Society*, **24**, 363–9.

44. Eichler, M. (1963). *Einführung in die Theorie der algebraischen Zahlen und Funktionen.* Birkhäuser, Basel.

45. Ellison, W.J. (1971). Waring's problem. *American Mathematical Monthly*, **78**, 10–36.

46. Estermann, T. (1952). *Introduction to modern prime number theory.* Cambridge University Press.

47. Goldstein, L.J. (1971). *Analytic number theory.* Prentice-Hall, Englewood Cliffs, NJ.

48. Haar, A. (1933). Der Massbegriff in der Theorie der kontinuierlichen Gruppen. *Annals of Mathematics*, **34**, 147–69.

49. Halberstam, H. and Richert, H.E. (1947). *Sieve methods.* Academic Press, London.

50. Halberstam, H. and Roth, K.F. (1966). *Sequences.* Clarendon Press, Oxford.

51. Hardy, G.H. (1918). On the representation of a number as the sum of any number of squares, and in particular of five or seven. *Proceedings of the National Academy of Sciences*, **4**, 189–93 = (1966). *The collected papers of G.H. Hardy,* Vol. I, pp.340–4. Oxford University Press.

52. Hardy, G.H. (1920). On the representation of a number as the sum of any number of squares, and in particular of five. *Transactions of the American Mathematical Society*, **21**, 255–84 = (1966). *The collected papers of G.H. Hardy,* Vol. I, pp.345–74. Oxford University Press.

53. Hardy, G.H. and Littlewood, J.E. (1920). Some problems of "partitio numerorum"; I: A new solution to Waring's problem. *Göttinger Nachrichten*, **5**, 33–54 = (1966). *The collected papers of G.H. Hardy*, Vol. I, pp.405–26. Oxford University Press.

54. Hardy, G.H. and Littlewood, J.E. (1921). Some problems of "partitio numerorum"; II: Proof that every large number is the sum of at most 21 biquadrates. *Mathematische Zeitschrift*, **9**, 14–17 = (1966). *The*

collected papers of G.H. Hardy, Vol. I, pp.427–40. Oxford University Press.

55. Hardy, G.H. and Littlewood, J.E. (1922a). Some problems of "partitio numerorum"; III: On the expression of a number as a sum of primes. *Acta Mathematica*, **44**, 1–70 = (1966). *The collected papers of G.H. Hardy*, Vol. I, pp.561–630. Oxford University Press.

56. Hardy, G.H. and Littlewood, J.E. (1922b). Some problems of "partitio numerorum"; IV: The singular series in Waring's problem and the value of the number $G(k)$. *Mathematische Zeitschrift*, **12**, 161–88 = (1966). *The collected papers of G.H. Hardy*, Vol. I, pp.441–68. Oxford University Press.

57. Hardy, G.H. and Littlewood, J.E. (1924). Some problems of "partitio numerorum"; V: A further contribution to the study of Goldbach's problem. *Proceedings of the London Mathematical Society (2)*, **22**, 46–56 = (1966). *The collected papers of G.H. Hardy*, Vol. I, pp.632–42. Oxford University Press.

58. Hardy, G.H. and Littlewood, J.E. (1925). Some problems of "partitio numerorum"; VI: Further researches in Waring's problem. *Mathematische Zeitschrift*, **23**, 1–37 = (1966). *The collected papers of G.H. Hardy*, Vol. I, pp.469–505. Oxford University Press.

59. Hardy, G.H. and Littlewood, J.E. (1928). Some problems of "partitio numerorum"; VIII: The number $\Gamma(k)$ in Waring's problem. *Proceedings of the London Mathematical Society (2)*, **28**, 518–42 = (1966). *The collected papers of G.H. Hardy*, Vol. I, pp.506–30. Oxford University Press.

60. Hardy, G.H. and Ramanujan, S. (1918). Asymptotic formulae in combinatory analysis. *Proceedings of the London Mathematical Society (2)*, **17**, 75–115.

61. Hardy, G.H. and Wright, E.M. (1979). *An introduction to the theory of numbers*, (5th edn). Oxford University Press.

62. Hayes, D.R. (1963a). A polynomial analog of the Goldbach conjecture. *Bulletin of the American Mathematical Society*, **69**, 115–6.

63. Hayes, D.R. (1963b). Correction to a polynomial analog of the Goldbach conjecture. *Bulletin of the American Mathematical Society*, **69**, 493.

64. Hayes, D.R. (1965). The distribution of irreducibles in $GF[q, x]$. *Transactions of the American Mathematical Society*, **117**, 101–27.

65. Hayes, D.R. (1966). The expression of a polynomial as a sum of three irreducibles. *Acta Arithmetica*, **11**, 461–88.

66. Hayes, D.R. (1972). Adèlic analysis in additive number theory. *Proceedings of the 1972 Number Theory Conference*, University of Colorado, Boulder, pp.106–7.

67. Hayes, D.R. (1974). Explicit class field theory for rational function fields. *Transactions of the American Mathematical Society*, **189**, 77–91.

68. Hayes, D.R. and Nutt, M.D. (1982). Reflective functions on p-adic fields. *Acta Arithmetica*, **40**, 229–48.

69. Hilbert, D. (1909). Beweis für die Darstellbarkeit der ganzen Zahlen durch eine feste Anzahl n-ter Potenzen. *Mathematische Annalen*, **67**, 281–300.

70. Knopfmacher, J. (1975). *Abstract analytic number theory*. North-Holland, Amsterdam.

71. Knopfmacher, J. (1979). *Analytic arithmetic of algebraic function fields*, Lecture Notes in Pure and Applied Mathematics, Vol. 50. Dekker, New York.

72. Kornblum, H. (1919). Über die Primfunktionen in einer arithmetischen Progression. *Mathematische Zeitschrift*, **5**, 100–11.

73. Körner, O. (1960). Übertragung des Goldbach–Vinogradovschen Satzes auf reell quadratische Zahlkörper. *Mathematische Annalen*, **141**, 343–66.

74. Kubota, R.M. (1974). *Waring's problem for* $\mathbf{F}_q[x]$. Dissertationes Mathematicae (Rozprawy Matematyczne), No. 117. Mathematical Institute of the Polish Academy of Sciences, Warsaw.

75. Landau, E. (1947). *Vorlesungen über Zahlentheorie,* Vol. I, Part 2. Chelsea, New York.

76. Lang, S. (1970). *Algebraic number theory.* Addison-Wesley, Reading, MA.

77. Leahey, W. (1967). Sums of squares of polynomials with coefficients in a finite field. *American Mathematical Monthly*, **74**, 816–19.

78. Lidl, R. and Neiderreiter, N. (1983). Finite fields. In *Encyclopedia of mathematics and its applications* (ed. G-C. Rota), Vol. 20. Addison-Wesley, Reading, MA.

79. Loomis, L. (1953). *An introduction to abstract harmonic analysis.* Van Nostrand, Princeton, NJ.

80. Mackey, G.W. (1978). *Unitary group representations in physics, probability, and number theory.* Benjamin/Cummings Publishing Company, Reading, MA.

81. Mars, J.G.M. (1973). Sur l'approximation du nombre de solutions de certaines equations Diophantiennes. *Annales Scientifiques de l'École Normale Supérieure*, **4**, 357–88.

82. Mordell, L.J. (1933). The number of solutions of some congruences. *Mathematische Zeitschrift*, **37**, 207.

83. Nachbin, L. (1965). *The Haar integral.* Van Nostrand, Princeton, NJ.

84. Paley, R.E.A.C. (1933). Theorems on polynomials in a Galois field. *Quarterly Journal of Mathematics*, **4**, 52–63.

85. Paley, R.E.A.C. and Wiener, N. (1934). *Fourier transforms.* American Mathematical Society Colloquium Publications XIX, New York.

86. Pfander, G. (1981). *Das Waring–Goldbach–Problem im Polynomring über einem endlichen Körper.* Abteilung für Mathematik, No. 4. Universität Ulm.

87. Ramanujan, C.P. (1963). Sums of m-th powers in p-adic rings. *Mathematika*, **10**, 137–46.

88. Rédei, L. (1946). Zur Theorie der Gleichungen in endlichen Körpern. *Acta Univ. Szeged*, **11**, 63–70.

89. Rhin, G. (1972). *Répartition modulo 1 dans un corps de séries formelles sur un corps fini.* Dissertationes Mathematicae (Rozprawy Matematyczne), No. 95. Mathematical Institute of the Polish Academy of Sciences, Warsaw.

90. Samuel, P. (1970). *Algebraic theory of numbers.* Houghton Mifflin, Boston.

91. Schmidt, W.M. (1976). *Equations over finite fields, an elementary approach*, Lecture Notes in Mathematics 536. Springer-Verlag, Berlin.

92. Schwarz, S. (1948a). On Waring's problem for finite fields. *Quarterly Journal of Mathematics*, **19**, 123–8.

93. Schwarz, S. (1948b). On the equation $a_1x_1^k + a_2x_2^k + \cdots + a_kx_k^k + b = 0$ in finite fields. *Quarterly Journal of Mathematics*, **19**, 160–3.

94. Serre, J.P. (1965). *Lie algebras and Lie groups.* Benjamin, New York.

95. Serre, J.P. (1979). *Local fields.* Springer-Verlag, New York.

96. Serre, J.P. (1981). Quelques applications du théorème de densité de Chebotarev. In *Publications Mathématique, n° 54*, pp.123-201. Institut des Hautes Étude Scientifiques, Bures-sur-Yvette.

97. Serre, J.P. (1988). *Algebraic groups and class fields.* Springer-Verlag, New York.

98. Shafarevich, I.R. (1977). *Basic algebraic geometry.* Springer-Verlag, Berlin.

99. Shuck, J.W. (1969). Calculus on p-adic manifolds, with applications to number theory. Unpublished Ph.D. thesis, Northeastern University, Boston.

100. Siegel, C.L. (1935). Über die analytische Theorie der quadratischen Formen. *Annals of Mathematics*, **36**, 527–606 = (1966). *Gesammelte Abhandlungen*, Band I, pp.326–405. Springer-Verlag, Berlin.

101. Siegel, C.L. (1944). Generalization of Waring's problem to algebraic number fields. *American Journal of Mathematics*, **66**, 122–36 = (1966). *Gesammelte Abhandlungen*, Band II, pp.406–20. Springer-Verlag, Berlin.

102. Siegel, C.L. (1951). Indefinite quadratische Formen und Funktionentheorie I. *Mathematische Annalen*, **124**, 17–54 = (1966). *Gesammelte Abhandlungen*, Band III, pp.105–42. Springer-Verlag, Berlin.

103. Siegel, C.L. (1952). Indefinite quadratische Formen und Funktionentheorie II. *Mathematische Annalen*, **124**, 364–87 = (1966). *Gesammelte Abhandlungen*, Band III, pp.154–77. Springer-Verlag, Berlin.

104. Tate, J.T. (1950). Fourier analysis in number fields and Hecke's zeta-functions. In *Algebraic number theory* (ed. J.W.S. Cassels and A. Frölich), pp.305–47. Thompson Book Company, Washington, DC.

105. Tatuzawa, T. (1955). Additive prime number theory in an algebraic number field. *Journal of the Mathematical Society of Japan*, **7**, 409–23.

106. Tornheim, L. (1938). Sums of n-th powers in fields of prime characteristic. *Duke Mathematical Journal*, **4**, 359–62.

107. Uchiyama, S. (1955). Sur les polynomes irréductibles dans un corps fini, II. *Proceedings of the Japan Academy of Science*, **31**, 267–9.

108. Val'fiš, A.Z. (1952). On the representation of numbers by sums of squares: asymptotic formulas. *Uspehi Matematičeskih Nauk (N.S.)*, **7**, no.6(52), 97–178 = (1956). *Translations of the American Mathematical Society (Series 2)*, **3**, 163–248.

109. Vaughan, R.C. (1979). A survey of some important problems in additive number theory. *Société Mathématique de France: Astérisque*, **61**, 213–22.

110. Vaughan, R.C. (1981). *The Hardy–Littlewood method.* Cambridge University Press.

111. Vaserstein, L.N. (1987a). Waring's problem for algebras over fields. *Journal of Number Theory*, **26**, 286–98.

112. Vaserstein, L.N. (1987b). Waring's problem for commutative rings. *Journal of Number Theory*, **26**, 299–307.

113. Vaserstein, L.N. (1991). Sums of cubes in polynomial rings. *Mathematics of Computation*, **56**, 349–57.

114. Vinogradov, I.M. (1937). Representation of an odd number as a sum of three primes. *Comptes Rendus (Doklady) de l'Académie des Sciences de l'URSS*, **15**, 191–294.

115. Vinogradov, I.M. (1952). *The method of trigonometric sums in the theory of numbers.* Interscience Publishers, London.

116. Webb, W.A. (1973a). Waring's problem in $GF[q,x]$. *Acta Arithmetica*, **22**, 207–20.

117. Webb, W.A. (1973b). On the representation of polynomials over finite fields as sums of powers and irreducibles. *Rocky Mountain Journal of Mathematics*, **3**, 23–9.

118. Webb, W.A. (1973c). Numerical results for Waring's problem in $GF[q,x]$. *Mathematics of Computation*, **27**, 193–6.

119. Weil, A. (1948). *Sur les courbes algébriques et les variétés qui s'en déduisent.* Hermann, Paris.

120. Weil, A. (1962). Sur la théorie des formes quadratiques. *Colloque sur la Théorie des Groupes Algébriques, C.B.R.M.* Brussels, pp.9–22 = (1979). *Collected papers*, Vol. II, pp.471–84. Springer-Verlag, New York.

121. Weil, A. (1964). Sur certaines groupes d'opérateurs unitaires. *Acta Mathematica*, **111**, 143–211 = (1979). *Collected papers*, Vol. III, pp.1–69. Springer-Verlag, New York.

122. Weil, A. (1965). Sur la formula de Siegel dans la théorie des groupes classiques. *Acta Mathematica*, **113**, 1–87 = (1979). *Collected papers*, Vol. III, pp.71–157. Springer-Verlag, New York.

123. Weil, A. (1967). *Basic number theory*. Springer-Verlag, New York.

124. Weyl, H. (1916). Über die Gleichverteilung der Zahlen mod eins. *Mathematische Annalen*, **77**, 313–52.

125. Wiener, N. (1933). R. E. A. C. Paley—In Memoriam. *Bulletin of the American Mathematical Society*, **39**, 476.

Index